WORLD BANK WORKING PAPER NO. 40

Power's Promise

Electricity Reforms in Eastern Europe and Central Asia

Edited by Julian Lampietti

THE WORLD BANK
Washington, D.C.

Copyright © 2004
The International Bank for Reconstruction and Development / The World Bank
1818 H Street, N.W.
Washington, D.C. 20433, U.S.A.
All rights reserved
Manufactured in the United States of America
First printing: June 2004

 printed on recycled paper

1 2 3 4 06 05 04

World Bank Working Papers are published to communicate the results of the Bank's work to the development community with the least possible delay. The manuscript of this paper therefore has not been prepared in accordance with the procedures appropriate to formally-edited texts. Some sources cited in this paper may be informal documents that are not readily available.

The findings, interpretations, and conclusions expressed in this paper are entirely those of the author(s) and do not necessarily reflect the views of the Board of Executive Directors of the World Bank or the governments they represent. The World Bank does not guarantee the accuracy of the data included in this work. The boundaries, colors, denominations, and other information shown on any map in this work do not imply on the part of the World Bank any judgment of the legal status of any territory or the endorsement or acceptance of such boundaries.

The material in this publication is copyrighted. The World Bank encourages dissemination of its work and normally will grant permission for use.

Permission to photocopy items for internal or personal use, for the internal or personal use of specific clients, or for educational classroom use, is granted by the World Bank, provided that the appropriate fee is paid. Please contact the Copyright Clearance Center before photocopying items.
Copyright Clearance Center, Inc.
222 Rosewood Drive
Danvers, MA 01923, U.S.A.
Tel: 978-750-8400 • Fax: 978-750-4470.

For permission to reprint individual articles or chapters, please fax your request with complete information to the Republication Department, Copyright Clearance Center, fax 978-750-4470.
All other queries on rights and licenses should be addressed to the World Bank at the address above, or faxed to 202-522-2422.

ISBN: 0-8213-5900-2
eISBN: 0-8213-5901-0
ISSN: 1726-5878

Julian Lampietti is Senior Social Development Economist in Environmentally and Socially Sustainable Development Sector Unit for the Europe and Central Asia Region at the World Bank.

Cover Photo by Otar Jangvaladze. This photo reflects the current electricity situation in Georgia.

Library of Congress Cataloging-in-Publication Data has been requested.

TABLE OF CONTENTS

Foreword vii
Abstract ix
Acknowledgments xi
Acronyms and Abbreviations xiii
Executive Summary xv

1. **The Promise of Reform** 1
   ECA is Different 1
   Power Sector Reforms are Urgent 2
   Proper Sequencing is Critical 3
   Net Political Benefits Explain Variation in Progress 4
   Taking Care of Institutions in the Energy Reform Roadmap 6
   Identifying Outcomes is Difficult 6

2. **Getting the Utilities off the Books** 9
   Quantifying the Sector Losses 9
   Deficits Declined for Different Reasons 11
   Relationship Between the Electricity Sector Deficit and the Fiscal Deficit ... 12
   Reform Savings Did Not Go to Social Spending 13
   Conclusions 15

3. **Creating More Efficient Companies** 17
   Improve Transparency and Accountability 17
   Revenue per kilowatt hour Rose 18
   Generation Costs Declined 19
   System Losses Held Steady 20
   Collection Rates Varied 20
   Mixed Results on Operational Efficiency 21
   Focus on Service Quality 23
   Regional Companies Take on the Challenge 23
   Conclusions 25

4. **Maintaining Power to the Poor** 27
   Household Energy Use Patterns Differ 28
   Tariffs Rose 29
   The Burden Increased 30
   Consumption is Low 32
   Gas May Be Filling the Gap 33
   Enforcement is Necessary 35
   Searching for Better Transfers 36
   Conclusions 39

5. **Local Versus Global Environmental Benefits** 41
   Did the Reforms Achieve Environmental Benefits? 41
   There Were Unintended Environmental Costs 44
   Damage from Dirty Fuel Use May Be Large 44
   Conclusions 47

Annexes **49**
    Annex A: Overview of the Reform Process in Eight ECA Countries 49
    Annex B: Proceeds from Privatization of Electric Utility Companies 53
    Annex C: Tariff Losses, Commercial and Collection Losses, as Share of Total Losses 55
    Annex D: Fiscal Balance and Electricity Sector Financial Deficit (Million USD
              and share of GDP) 57
    Annex E: Efficiency Indicators 59
    Annex F: More on the Methodology for Estimating Health Effects 63
    Annex G: Base Emission Factors 65
    Annex H: More on Factors Leading to Low Contribution of Power Sector
              Toward Health Damages 67
    Annex I: Changes in Generation Mix in the Past Decade 69
    Annex J: Proposed Energy Issues to be Addressed and Sample Questions
              in LSMS/HBS Surveys 71

References **73**

TABLE OF FIGURES

Figure 1.1: Suggested Sequencing of Power Sector Reforms in ECA 4
Figure 2.1: The Electricity Sector's Losses Declined, 1993–2001 11
Figure 2.2: Power Sector Losses are Weakly Correlated with Fiscal Deficits, 1995–2000 13
Figure 2.3: Poor Countries Spend Less on Poor People (1997–2000) 14
Figure 2.4: Social Spending on Education, Health and Social Assistance Did Not Increase
          (as percent of GDP) 15
Figure 3.1: Nominal Revenues Increased Marginally, 1991–2001 19
Figure 3.2: System Losses Stayed Stable, 1990–2002 20
Figure 3.3: Collection Rates Differ by Country, 1990–2002 21
Figure 4.1: Residential Electricity Tariff—Index of CPI Adjusted Real Tariffs 30
Figure 4.2: Enough for Three Light Bulbs and a Refrigerator 33
Figure 4.3: Electricity Tariffs are Higher Than Those for Gas, 1992–2002 35
Figure 4.4: Burden of Arrears is the Same for the Poor and Non-poor, 2000–2002 36
Figure 4.5: Alternative Subsidy Scheme 37
Figure 5.1: Fuel Required to Produce 1 Megawatt-hour of Electricity, 1992–98 42
Figure 5.2: Electricity Contributes in Small Measure to Health Damages 45

TABLE OF TABLES

Table 1.1: ECA is Different 2
Table 1.2: Timeline of Reforms in the Electricity Sector in ECA 4
Table 1.3: Independence of Regulatory Institutions: Paper and Practice 6
Table 1.4: Investment Climate Varies Among Reformers 7
Table 3.1: Suggested Indicators of Production Efficiency 18
Table 3.2: Reported Generation Costs Went Down, 1990–2002 (cents/kilowatt hour) 19
Table 3.3: Aggregate Impact of Reform on Collection Rates in Tbilisi 22

Table 3.4: Service Quality Improved 24
Table 4.1: Urban Network Energy Use in ECA, (percent of households) 28
Table 4.2: Urban Non-network Energy Use in ECA (percent of households) 29
Table 4.3: Shares of Spending on Electricity Went Up, 1993–2002 31
Table 4.4: Consumer Surplus Fell 32
Table 4.5: Electricity Expenditure as a Share of Total Energy Expenditure, 1993–2002 34
Table 4.6: Simulation of Subsidy Cost-effectiveness for Tbilisi, Georgia 38
Table 5.1: Exposure to Indoor Air Pollution is High 46

TABLE OF BOXES

Box 2.1: Untangling the Quasi-fiscal Deficit 10
Box 2.2: Calculating the Implicit Subsidies (Sector Losses) 10
Box 2.3: Social Spending in Latin America Increased Since the 1990s 16
Box 3.1: Private Sector Improving Household Collections in Georgia 21
Box 3.2: Opportunistic Behavior by the Private Sector? 25
Box 4.1: Simulation of Alternative Subsidy 37
Box 5.1: Reform Measures Expected to Result in Environmental Quality Improvements ... 42
Box 5.2: Estimating the Power Sector's Contribution to Air Pollution
and Health Damage 43
Box 5.3: Methodology for Calculating Damages from Indoor Air Pollution 46

FOREWORD

Energy sector reforms remain among the most controversial development issues in transition economies, as these countries continue to tread the path toward sustainable growth. The legacy of central planning left the sector highly centralized, vertically integrated, and often inefficient and deeply in debt. The reforms promised improved fiscal balances, more efficient resource use, better consumer service, and environmental benefits. This study ventures beyond the immediate fiscal and efficiency impacts of electricity reforms to include the social and environmental impacts. At one level, it can be viewed as a cost-benefit analysis of power sector reforms in transition economies.

The countries in this study—Armenia, Azerbaijan, Georgia, Hungary, Kazakhstan, Moldova, and Poland—started reforming their electricity sectors in the 1990s. Different starting points with respect to income and political and institutional environment influenced sector reform implementation; and the adjustment path was a function of both internal and external political and economic pressures.

The study is timely because increasing emphasis is being placed on demonstrating the results of our work. It points to the need to improve our understanding of the linkage between sector reform and fiscal outcomes and how we can better monitoring of the fiscal and efficiency impacts of reform. It also suggests how we can design better strategies to mitigate the impact of reform on the poor and the environment. Ultimately the results of this study and the analytic framework it presents should be helpful for countries interested in assessing the full set of tradeoffs between equity and efficiency in power sector reform.

Laura Tuck
Sector Director
Environmentally and Socially Sustainable Development
Europe and Central Asia

Abstract

This study analyzes the fiscal, efficiency, social, and environmental impact of power sector reforms in seven countries in the ECA region. It finds sector deficits have been falling over the last decade and that the savings from lower sector deficits did not translate into higher social spending. More emphasis must be placed on monitoring deficits and tailoring policy reform to country specific circumstances. The impact of reform on utility efficiency, as measured by the cost of generation, system loss collections, and operational efficiency, is ambiguous. While overall revenue per kilowatt hour increased in almost all countries, problems continue with losses, collection rates, and staffing. In terms of social impacts, electricity spending as a share of income increased, especially for the poor, while consumption stayed the same. In terms of environmental impacts, reforms did slightly improve energy efficiency in power plants though this has little direct impact on human health because the electricity sector's share of the total health damage from air pollution is negligible. Several lessons emerge from the analysis. Undertaking simple *ex ante* simulations of reform impacts will allow better identification of potential reform benefits and costs. Placing more emphasis on outcome-based indicators of service quality would help ensure that future operations produce the intended end-user benefits. In many cases, tariff increases can and should be explicitly timed to coincide with service quality improvements. Yet, this may not be always possible. Where it is not, the adverse impact of tariff increases, especially for low-income consumers, should be mitigated by improving access to and efficiency in the use of clean alternatives.

Acknowledgments

A multi-sectoral team from ECSSD, ECSIE, and ECSPE prepared this report under the supervision and guidance of Alexandre Marc (Sector Manager, ECSSD) and Lee Travers (Sector Manager, ECSIE). Julian A. Lampietti (ECSSD) managed the task. Task team members in alphabetical order include Sudeshna Ghosh Banerjee (ECSSD), Julia Bucknall (ECSSD), Peter Dewees (ECSSD), Jane Ebinger (ECSIE), Irina Klytchnikova (ECSSD), Taras Pushak (ECSSD), Gevorg Sargsyan (ECSIE), Sergei Shatalov (ECSPE), Maria Shkaratan (ECSSD), and Katelijn Van den Berg (ECSSD).

The team received support from local consultants including V. Valiyev in Azerbaijan, N. Nadiradze in Georgia, E. Nersisyan and A. Marjanyan in Armenia, S. Katyshev and G. Mandrovskaya in Kazakhstan, L. Lengyel in Hungary, and R. Gillecki from ARE in Poland. We are thankful to Ani Balabanyan for overseeing the work of local consultants in Armenia. Thanks are due to Maureen Cropper (DECRG), Kseniya Lvovsky (SASES), and László Varró of the Hungarian Energy Regulatory Office. We also thank Marcin Sasin in the Warsaw office of World Bank and Zsuzsa Szabo of Hungary National Statistical Office for help with household surveys. We are grateful to Bruce Ross-Larson of Communications Development Inc. for editing this report.

Peer review comments were provided by Mamta Murthi (ECAVP), Gary Stuggins (EWDEN), Philippe Durand (LCSFE), Vivien Foster (LCSFP), Kseniya Lvovsky (SASES), and Jan Bojo (ENV). Navroz Dubash (World Resources Institute) and Ian Walker (Economic Adviser, Government of Honduras) provided comments on the concept note. Our partners in the local World Bank offices, Farid Mamedov in Azerbaijan, Pawel Kaminski in Poland, Gayane Minasyan in Armenia, Ilia Kvitaishvili in Georgia, provided comments and support throughout the study. Financial support from the Poverty Window of the Norwegian Trust Fund for Environment and Socially Sustainable Development and ECA Vice Presidency is gratefully acknowledged. We particularly want to thank Giovanna Prennushi and Ignacio Fiestas of the Norwegian Trust Fund for Environment and Socially Sustainable Development for their continued support of this project.

Acronyms and Abbreviations

CHP	Combined Heat and Power
CIS	Commonwealth of Independent States
CPI	Consumer Price Index
DALY	Disability Adjusted Life Years
DECRG	Development Research Group
EBRD	European Bank for Reconstruction and Development
ECA	Europe and Central Asia
ECAVP	Office of the Regional Vice President, Europe and Central Asia Region
ECSIE	ECA Infrastructure and Energy, The World Bank
ECSPE	Poverty Reduction and Economic Management unit, Europe and Central Asia Region
ECSSD	Environment and Socially Sustainable Development unit, Europe and Central Asia Region
EU	European Union
EWDEN	Energy Anchor
GDP	Gross domestic product
HBS	Household Budget Survey
ICRG	International Country Risk Guide
IMF	International Monetary Fund
kWh	Kilowatt Hour
LCSFE	Energy Cluster, Latin America and Caribbean Region
LCSFP	Finance, Private Sector and Infrastructure unit, Latin America and Caribbean Region
LPG	Liquefied Petroleum Gas
LSMS	Living Standards Measurement Surveys
SASES	Environment and Social unit, South Asia Region
UN	United Nations
UNEP	United Nations Environment Program
USAID	United States Agency for International Development
VAT	Value Added Tax
WHAP	Winter Heat Assistance Program
WHO	World Health Organization

Executive Summary

The socialist legacy left a power infrastructure in Europe and Central Asia (ECA) that was equitable yet extremely inefficient. To speed the region's economies through the transition, substantial emphasis was placed on reforms to increase efficiency in the sector. The promises of reform are improved fiscal balances, more efficient resource use, better consumer service, and environmental benefits. Ultimately, the deadweight loss is eliminated, and public resources are freed up for more productive investments.

This study, intended to inform the design of power sector reforms in the region, identifies what has worked in promoting both equity and efficiency. It systematically brings together a broad range of cross-country indicators for the most aggressive reformers in the region over the last decade: Armenia, Azerbaijan, Georgia, Hungary, Kazakhstan, and Poland.

Did the outcomes materialize as intended? While it may be too early to say, all of the chapters point to serious data deficiencies. In general it is possible to conclude that better design of monitoring and data collection systems is required. In addition to enabling the assessment of reform impacts, those systems would improve program design by producing the data to allow simulation of the full range of policy options of interest to the client.

Getting the Utilities Off the Books

The legacy of central planning was a power sector contributing to unsustainable government deficits. So getting the power utilities off the government's books is the key to reducing fiscal budget deficit and promoting macroeconomic stabilization and growth.

The sector losses have been falling over the last decade, with tariffs as the key determinant. Yet, more emphasis should have been put on monitoring those losses. Country-specific cost-recovery tariffs can generate more precise estimates of these losses. This work is under way in a separate regional initiative.

Despite the decline in sector losses, the impact of reforms on the fiscal deficit is ambiguous. Why? Because of a combination of factors, including the lack of data, very complex financial and budget flows, and different adjustment paths. Only a detailed analysis of the quasi-fiscal accounts would lead to a definitive conclusion about the impact of the reduction in sector losses on the fiscal deficit. On balance, we conclude that assessment of budgetary impact of the reform can be improved by carefully examining the relative share of the energy and non-energy subsectors. Positive outcomes are expected in countries where a higher share of GDP is from the energy sector. However, institutional reform in these countries will also be the most difficult.

The savings from lower sector losses did not translate into higher social spending. It may be that savings from reforms went to reducing prior sector obligations. The share of spending on social assistance, education, and health stayed the same or fell. Aggregate social assistance data show that these costs are much smaller than the savings generated by a reduced losses. This implies that a larger share of savings can presumably go toward alleviating the blow from tariff increases. It does not imply that financing consumption (rather than more productive investments in health and education) is the best way to do this.

Creating More Efficient Companies

Electricity reforms are expected to improve the production efficiency of the power sector, translating into cost savings and service quality improvements for end users. They are also designed to make the sector financially sustainable by increasing the efficiency of resource allocation and the cost effectiveness of sector investments. The efficiency gains come from a profit motive—created by more competition or transparent regulation.

While the legacy of central planning suggests large potential production efficiency gains, the data collected here on the cost of generation, system loss collections, and operational efficiency are ambiguous. Despite continuing problems with losses, collection rates, and staffing, overall revenue per kilowatt hour increased in almost all countries. The beneficiary of the additional revenues from declining generation costs and rising retail tariffs is not so clear. Limited aggregate data suggest that service quality improved in a number of capital cities.

Simple *ex-ante* simulations would allow better quantification of potential reform benefits and costs. Placing more emphasis on outcome-based indicators of service quality would help ensure that future operations produce the intended end-user benefits. Both the private and the public sector can improve utility performance, suggesting that a broad range of contract arrangements can produce the desired outcomes. Finally, a number of local investors with both region and sector experience are filling investment needs, suggesting alternatives to the traditional definition of a strategic investor.

Maintaining Power to the Poor

Unlike other regions, the socialist system gave almost all households access to reliable, inexpensive electricity. So the welfare gains from increased access—one of the most immediate and tangible benefits of power sector reforms—is not a consideration in most ECA countries. The welfare gains come from improvements in service quality, again reinforcing the need for reform to emphasize outcome-based indicators of service quality.

Without disaggregated baseline data on service quality, reform appears closely linked to a fall in welfare. Electricity spending as a share of income increased, especially for the poor, while consumption stayed the same. Several lessons emerge. In many cases, tariff increases can and should be explicitly timed to coincide with service quality improvements. Yet, this may not always be possible. Where it is not, the adverse impact of tariff increases, especially for low-income consumers, should be mitigated by improving access to and efficiency in the use of clean alternatives. In some locations, especially urban areas where households heat with electricity, natural gas may be a viable substitute.

There are serious problems of self-reported electricity data (as well as energy expenditure data in general) collected in the traditional poverty monitoring surveys such as the Living Standard Measurement Studies and Household Budget Survey. The reason is that the questions are confounded by recall error, under and over reporting, and the presence of arrears, making it impossible to identify current and historical consumption. This also raises general concerns about the treatment of electricity consumption, an important part of household budgets in many countries, in welfare aggregates. Where actual household data on consumption have been collected (directly from the utilities), consumption is very low and sufficient only to satisfy basic subsistence needs. In countries with very low consumption, demand for electricity is relatively inelastic, suggesting that there may be large welfare losses associated with future tariff increases.

There may be substantial positive social benefits associated with access to electricity, suggesting an important continuing role for the public sector. This report is agnostic on the empirical evidence of the effectiveness of alternative instruments that mitigate the blow of tariff increases on the poor, such as income transfers and lifelines. It does, however, advocate helping clients calculate the social and fiscal implications of a full set of alternative mitigating strategies so that they can make fully informed choices.

Local versus Global Environmental Benefits

Power sector reforms are expected to produce environmental benefits. Higher production efficiency, new investment, and environmentally friendly technologies all contribute to lower fossil fuel consumption and lower emissions. This leads to better ambient air quality and presumably to better health outcomes for the local population.

Unfortunately, claims about improvements in ambient air quality are difficult to verify because monitoring programs were never established. Better monitoring of ambient environmental quality improvements is necessary in the future. For the long-term sustainability of reforms, it is critical that environmental impacts be systematically followed and recorded. It is appropriate to define mitigation measures for negative environmental impacts, such as those on health. Impact analysis should be on a national scale, not from a sector perspective.

Reforms did slightly improve energy efficiency in power plants. This matters for carbon dioxide emissions and global climate impact. In most cases this has little direct impact on human health since the electricity sector's share of the total health damage from air pollution is negligible.

It is also possible that reforms have damaged health because households switched to dirty fuels. Household fuel burning has a high impact on ambient urban air quality because households often live in densely populated areas in houses with low chimneys and no pollution abatement equipment. As noted in the previous section, a key solution is improving access to and efficiency in the use of clean alternatives. Survey data indicate that fewer households would use wood and coal if they had access to gas. Of course, in many countries the gas sector is also in need of reform before it can operate on a sustainable basis.

As in the previous section, household surveys do not reveal enough information about energy (and other utility) reforms. More data are needed to evaluate the impact of reforms on fuel switching, energy use, substitution effects, and health and social impacts. This calls for including questions about utilities in such surveys and developing models to help predict behavior under a variety of scenarios.

CHAPTER 1

THE PROMISE OF REFORM

In Europe and Central Asia (ECA), the socialist legacy left a power infrastructure that was equitable[1] yet extremely inefficient. To speed the region's economies through the transition, substantial emphasis has been placed on efficiency-increasing reforms in the power sector—coupled, where necessary, with redistribution through lump-sum transfers.

Unlike other regions where reforms promise increased access to electricity, one of the major objectives in ECA is to prevent further service quality deterioration. The long-run promise of power sector reforms in ECA includes improved fiscal balances, more efficient resource use, better consumer service, and environmental benefits as energy efficiency increases. Ultimately the deadweight loss is eliminated, and public resources are freed for more productive investments.

This report tries to determine if the reforms produced the intended outcomes. In a perfectly competitive economy, tradeoffs between equity and efficiency take place along a production frontier (Birdsall and Nellis 2003). In principle, ECA economies are well inside the production frontier, so a balanced reform strategy could offer potential for growth without sacrificing equity. The report attempts to identify what has worked in promoting both equity and efficiency, recognizing the importance of externalities, information asymmetries, rent-seeking behavior, and other attributes of imperfect markets. This study, intended for policymakers in the ECA region, informs the broader debate on impact of energy reforms.

ECA is Different

The starting point for energy reforms in ECA is different from the other regions, further complicating program design and implementation (Table 1.1). Incomes, measured by GDP per capita, are higher than in other parts of the world except Latin America. Similarly, other development indicators—such as infant mortality rate, illiteracy rate, and access to basic infrastructure such

---

1. In this context, equitable implies widespread access to electricity.

TABLE 1.1: ECA IS DIFFERENT

	GDP per Capita, PPP (Current International $)	Adult Illiteracy Rate	Infant Mortality Rate	Access to an Improved Water Source: % of Population	GDP per Unit of Energy Use: PPP $ per kg Oil Equivalent
East Asia & Pacific	3,730	14	35	76	na
Eastern Europe & Central Asia	6,271	3	31	91	2.3
Latin America & Caribbean	7,093	11	30	86	6.1
Middle East & North Africa	5,188	36	45	88	3.8
South Asia	2,236	45	72	84	5.5
Sub-Saharan Africa	1,647	39	106	54	2.9

Note: All the figures are for 2000.
Source: *World Development Indicators*, World Bank (2003).

as water—suggest that the situation in ECA is better. GDP per unit of energy use is substantially lower—due in part to the cold winters, the legacy of central planning, and substantial declines in household incomes following transition. The Soviet legacy resulted in publicly owned, vertically integrated, and highly centralized power infrastructure designed to provide reliable electricity to all households at little or no cost. Studies indicate that access to electricity is substantially higher than in other regions with similar incomes (Komives, Whittington, and Wu 2001; Clarke and Wallsten 2002). Electricity consumption levels are also thought to be higher than in other parts of the world, suggesting substantial potential efficiency gains (Cornille and Frankhauser 2002).

Power Sector Reforms are Urgent

Providing large numbers of consumers with reliable electricity at little or no cost requires substantial power infrastructure. When this infrastructure was built, energy prices in the Soviet Union were well below international prices. With the onset of transition and the end of central transfers, the sector's large investment needs were neglected because revenues were low, if positive at all. Low tariffs, high consumption levels, and low collections characterized the sector.

The net result was a power sector supported through fiscal and quasi-fiscal budget operations and asset depreciation, especially in countries of the former Soviet Union. In Armenia, Georgia, and Moldova the power sector deficit was among the largest items in the budget deficit. The energy (electricity and gas) sector's quasi-fiscal deficit has been estimated at 5 percent of GDP in Moldova in 1999, 3.5 percent in Armenia in 2001, and 2 percent in Georgia in 2000 (International Monetary Fund 2001a, 2001b, 2001c).

Thus, the opportunity cost of subsidizing the energy sector is large. As countries struggle to balance their budgets and reduce quasi-fiscal operations, transfers of this magnitude take limited funds away from other sectors. Public spending on health and education has fallen dramatically since the beginning of the transition. In Azerbaijan, Georgia, and Moldova spending on health is less than a quarter of what it was in the early 1990s—and in Kazakhstan, about half. Spending on education is a mere sixth of the level in the early 1990s in Armenia, and a third in Azerbaijan. As shares of GDP, total public expenditures on education, health, and social assistance and welfare remained stable or fell (Public Expenditure database, 2003).

Falling service quality, continuing lack of investment, and persistent sector deficits made the reforms urgent in the ECA region, and the donor community has been advising governments on how to undertake them since the early 1990s. The advice of the World Bank, detailed in the 1998 ECA energy sector strategy, includes raising prices to cost-recovery levels, metering and cutting off nonpaying customers, establishing predictable and transparent regulations, introducing competition in generation and supply, and selling industrial assets to private strategic investors (World Bank 1998). International donors supported energy conservation measures by encouraging price reforms, providing funding for improving energy efficiency, and providing advice on better targeting of transfers to poor households.

This study examines the reform outcomes for countries that have made significant progress. It evaluates their experience along four dimensions: fiscal and quasi-fiscal operations, productive efficiency, social welfare, and environment. Armenia, Azerbaijan[2], Georgia, Hungary, Kazakhstan, Moldova, and Poland were selected for this study because they have been among the most aggressive reformers over the last decade.[3] Before analyzing these outcomes in subsequent chapters, the rest of this chapter explores factors driving the sequence and pace of reform in these diverse economies.

Proper Sequencing is Critical

In most ECA countries the electricity sector reforms began almost a decade back, but privatization started only recently. Figure 1.1 presents the suggested sequence of reforms. However, each country followed its own path based on the domestic political and economic conditions. Some governments, like Hungary and Kazakhstan, have chosen to go as far as to open the generation subsector to foreign investment and privatize, while others deem the generation assets as strategic and retain public ownership.

A comprehensive review of the reform experience in ECA reveals that no universal recipe for reforms can fit the specific conditions of each country (Krishnaswamy and Stuggins, 2003). Hungary and Poland engaged in reform the earliest, followed by Kazakhstan, Moldova and Georgia (Table 1.2). Unlike all other countries, Hungary's electricity sector has been open to participation of foreign strategic investors from the outset of reforms, much like other "strategic" sectors in Hungary, such as banking. Because there was no problem with the payment discipline either before or after the beginning of reforms, distribution and generation companies were privatized simultaneously. Though Armenia entered late, significant progress was made in restructuring the sector. It made a number of abortive attempts to sell the loss-making distribution network to foreign strategic investors before the sale finally went through. In 2002, a controlling stake in its power distribution network was sold to an offshore company—Midland Resources Holding. It happened after substantial attempts by the government to reorganize the sector, improve service, and align prices. The pace of energy reforms was comparatively fast in Georgia, which has initiated all reform measures. Georgia's biggest foreign investor, U.S.-based AES Corporation, recently sold its assets after five years of operation and significant problems with tariffs, bill payments, and political disturbances. Kazakhstan privatized 80–90 percent of its generating capacity using asset sales and concessions in 1996, and it completed privatization of the rest of generation assets between 1999–2002 except for a hydro plant (Annex A provides a detailed description of the reform steps in each country).

The common theme in the experience of all countries is the central role of institutions and the regulator in a successful reform process. Another common theme is the sequencing of reform steps. Sequencing becomes crucially important in the presence of nonpayment. If generation assets are privatized before distribution, a poor collection rate will result in an undervaluation of

---

2. Azerbaijan placed its distribution networks under private management contract in 2002.
3. Moldova and Ukraine could also be considered in this group but were not included here because of data collection issues.

TABLE 1.2: TIMELINE OF REFORMS IN THE ELECTRICITY SECTOR IN ECA

	Date of Passage of Energy Law and Creation of an Independent Regulator	Corporatization and Unbundling	Privatization of Distribution	Privatization of Generation	Market Liberalization
Armenia	1997	1997	2002	none	none
Azerbaijan	1998	1996	2001–2003[c]	none	none
Georgia	1997	1999–2000	1998	2000	1999–2000
Hungary	1993–94	1993–94	1995	1996–97	2001
Kazakhstan	1998–99	1996	1996, 1999	1996, 1999–2002	b
Moldova	1998	1997	1999	None	None
Poland	1997	1993	Ongoing	None[a]	Ongoing

a. except for new entry of private strategic investors
b. large customers and generators are allowed to enter in bilateral contracts. In 2001 the government set up KOREM (Kazakh market operator for electric energy and capacity) to organize the spot and 'day ahead' markets.
c. concession contracts

Source: Adapted from Krishnaswamy and Stuggins, 2003.

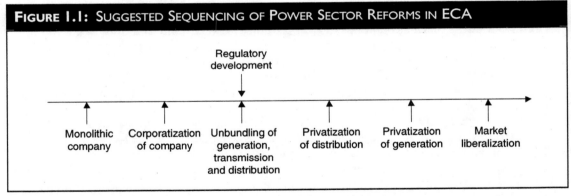

Source: European Bank for Reconstruction and Development (2001).

the generation assets, or a buyer may not be found at all. Sector unbundling becomes difficult when nonpayment is pervasive. One of the key lessons from the reform experience is that during the period of economic turmoil and pervasiveness of nonpayment, it is prudent to focus on the stabilization efforts, removal of legal, political and attitudinal constraints to denial of service to defaulters, and on other measures to improve collections and recover arrears. When these preconditions were not present, the reform progress was difficult and associated with social upheaval.

Net Political Benefits Explain Variation in Progress

Political, institutional, and macroeconomic conditions explain the variability in the pace of reforms and their outcomes. The decision to reform is determined by the net political benefits that accrue to policymakers—the difference between the benefits and costs of sector reform.

There is a mismatch between short-term costs and long-term benefits because costs are usually immediate and benefits accrue in the future (Rodrik 1994, Banerjee and Munger 2003).

It is widely acknowledged that while the demise of communist regimes and pressing fiscal concerns were the major factors driving sector reforms in all the economies in the region, nonpayment, widespread economic dislocation, and decline of industrial output accompanied this transition. Country-specific conditions—such as energy endowment, European Union (EU) accession, and the accumulation of energy related debt—also determined the reform's timing and intensity. Therefore, the reforms in Kazakhstan and Azerbaijan are analyzed here in the context of resource dependence theory, Hungary and Poland in the context of EU accession, and Armenia, Georgia, and Moldova in the context of energy-related debt. Paying careful attention to these factors and their implications for the speed of governments acting in the initial stages of reform design may well improve long-term performance.

Resources as a barrier. Energy resources can either hurt or ease progress with sector reform. They can ease the reform process by providing the resources necessary to prevent fiscal crises, but they can also be a barrier to reform implementation when resource rents are appropriated by the ruling elites (Esamov and others, 2001). Indeed, energy-rich countries (Kazakhstan and Azerbaijan) and energy-poor countries (Armenia, Georgia, and Moldova) differed in their reaction to reform and the pace of alignment with the world energy prices (Saavalainen and ten Berge 2003). While the energy-exporting countries gained from a change in their terms of trade during the transition period, now able to export their energy resources at the higher world price, the poorer countries lost. The latter countries accumulated energy-related debt and did not have resources within the energy sector to mitigate the adverse social impact of reforms.

Energy resources may also have slowed the pace of reforms.[4] It has been shown that the ruling elites benefit from partial reforms in lucrative sectors, gain control of the regulatory process, and prevent the creation of a level playing field (Hellman 1998). As Saavalainen and ten Berge (2003) note, the symbiotic relationship between energy utilities and the government meant widespread corruption and rent-seeking in the pre-reform era.

Accession to the European Union. The prospect of EU accession has provided the impetus for fast-paced reforms, especially in developing a regulatory framework and unified gas and electricity markets. The EU accession countries will have to conform with the EU directive on power reforms, which include liberalizing markets, decoupling generation, transmission, and distribution, and establishing regulated third-party access for the power network. In privatization, Hungary is the leader in pursuing major electricity privatizations in the 1990s, with most electric utilities privatized and prices at the world market level. Poland, following the dissolution of the communist regime, embarked on an ambitious "economic transformation program" in 1990. However, the Polish government has been more careful than Hungary to allow entry of foreign investors in the energy sector, deemed by the government as "strategic."[5]

Energy-related debt. The need to reduce energy-related external debt has been a significant driver of reform in the heavily indebted Armenia, Georgia, and Moldova (Saavalainen and ten Berge 2003). General economic, social, and political malaise in these countries following transition made power sector reforms urgent. Conditional lending was used in an effort to create a stable macroeconomic environment. For Armenia, Georgia, and Moldova, donor conditionalities focused on cost-recovery tariffs, collections, independent regulation, restructuring, unbundling, and privatization. Performance was mixed. During 1993–2002 only 60 percent of IMF energy

---

4. For suggestive evidence on this in other regions particularly on the philosophical debate and empirical evidence on the inverse relationship between natural resource abundance and economic growth, see Sachs and Warner (1995).

5. For more information on Hungary and Poland's EU requirements, refer to *Hungary: On the road to the European Union* (A World Bank Country Study) and "Poland—Country economic memorandum: reform and growth on the road to the EU" (World Bank).

Table 1.3: Independence of Regulatory Institutions: Paper and Practice

	Armenia	Azerbaijan	Georgia	Hungary	Kazakhstan	Moldova	Poland
Separate regulator	√	√	√	√	√	√	√
Fixed-term appointment	√		√			√	√
Industry funding			√	√		√	√
Full tariff-setting power	√		√		√	√	
Transparency	√		√	√	√	√	√
Redress	√	√	√	√	√	√	√

Source: Kennedy (2003), adapted from the EBRD/World Bank survey of regulators.

conditions were implemented (primarily relating to foreign energy debt and categorical privileges). Recognizing this, in the past year the IMF and the World Bank decreased the number of conditions in all countries except Georgia (Saavalainen and ten Berge 2003).

Taking Care of Institutions in the Energy Reform Roadmap

The biggest challenge in the reform process has been creating the appropriate institutional environment. Before the transition the energy sector in the former Soviet Union was a monopoly under direct state control. One major objective in the reform process is to encourage private interest to promote capital investment and harness technical expertise. For this to succeed, the government has to be committed to reform and have the adequate institutional capacity in place. Beyond doubt, the role of regulator is paramount in creating a level playing field and plausible dispute-resolution system.

Most of the countries in this study have progressed with regulatory reform, and all of them have set up regulatory agencies (Table 1.3). However, two unexpected problems occurred at the early stages of the process. The first was a shortage of regulators with technical experience in the region and its politics. This also suggests a role for international aid agencies to step in to provide technical assistance for setting up a system of fair and accountable regulation. The second was (and is) continued political interference in the appointment of regulators and in funding. Even undertaking all the regulatory steps may not ensure successful outcomes, because it takes a while for regulation to become effective.

Based on a survey of international power investors in developing countries, Lamech and Saeed (2003) found that investors require "enforceable legal system and credible rules of the game." The governments should be willing and able to respect its commitments and contractual obligations. It may appear that they do: for instance, the Georgian government has always overtly supported AES. Even so, government intervention is highest for Armenia, Azerbaijan, Moldova, and Georgia, and Georgia also tops the list with its entry barriers (Table 1.4). The problems of the government are exacerbated by the difficulty in identifying and evaluating the outcomes of the process, important for ensuring credibility of the reforms and justifying the price rise to citizens.

Identifying Outcomes is Difficult

Studying reform outcomes is complicated because it requires a counterfactual—what would have happened without reforms? Creating this counterfactual to evaluate the performance of the entire electricity sector is difficult because it requires many assumptions—not only about a particular company,[6] but about the entire domestic political and economic situation (for example,

---

6. A methodology for this type of analysis is detailed in Galal and others (1994).

TABLE 1.4: INVESTMENT CLIMATE VARIES AMONG REFORMERS

	Armenia	Azerbaijan	Georgia	Hungary	Kazakhstan	Moldova	Poland
Degree of government intervention[a]	3	3	3	1	2	3	2
Entry barriers[b]	11	na	13	8	12	na	11
Degree of property rights protection[c]	3	4	4	2	4	3	2
Quality of regulation[d]	4	4	4	3	4	4	3
Index of political risk[e]	41	40	na	20	31	32	22
Index of judicial system efficiency[f]	2	2	2	>1	>1	1	>1
Index of corruption[g]	6	6	5	3	6	6	2

a. High values imply high levels of government intervention. This factor measures 'government's direct use of scarce resources for its own purposes and government's control of resources through ownership'. Ranges from '1' (government expenditure less than of equal to 15 percent of GDP) to '5' (Greater than 30 percent'. Source: Heritage Foundation (2002).

b. High values imply difficult to enter. Source: Number of Procedures Needed to Start a Business (Djankov et al., 1999). Data are for 1999.

c. High values imply weak protection. This factor measures the degree of protection to private property and how laws are enforced to protect private property. Ranges from '1' (very high protection) to '5' (very low protection) Source: Heritage Foundation (2002).

d. High values imply poor regulation. This factor measures the extent of difficulty in operating a business. It also examines the degree of corruption in government and whether regulations are imposed fairly in all businesses. Ranges from '1' (very low levels of regulation) to '5' (very high levels of regulation) Source: Heritage Foundation (2002).

e. High values imply greater risk. Source: ICRG (June 2002). [100-ICRG political risk score].

f. High values imply poor performance. Source EBRD (2001) (4-score on legal effectiveness).

g. High values imply greater corruption. Indicator of 'perception of corruption in civil service, the business interests of top policy makers, laws of financial disclosure, and conflict of interest, and anti-corruption activities'. Source: Nations in Transit (2001).

Source: Adapted from Lieberman and others 2003, World Bank (2003f).

alternative scenarios of economic growth, private investment, and political stability). The alternative of trying to compare outcomes before and after reform ignoring the counterfactual is also difficult because of the debate over the appropriate baseline (pre-transition or transition), the fact that reforms are dynamic, and the scarcity of pre-reform data. Therefore this paper analyzes performance trends over the last decade using data compiled from a number of sources including local consultant reports, reviews of project documents, Household Budget Surveys (HBS), and Living Standards Measurement Study (LSMS) surveys. There are substantial methodological challenges in identifying outcomes, particularly the social and environmental. The methods and issues raised here can be used in the new Poverty and Social Impact Analysis initiative in the Fund and the Bank to explicitly incorporate the findings into reform design.

Caution is required in drawing conclusions because reforms are a dynamic process and the countries examined are at different points in the reform process (Table 1.2). Hungary and Poland started the reforms as early as 1992, while most of the Commonwealth of Independent States (CIS) countries initiated them in 1995 or 1996. Of course, it has to be kept in mind that most of the CIS economies became countries only the beginning of the decade and tasks of nation-building took precedence. It may thus be too early to draw conclusions about outcomes where the process is not complete. In addition, the time elapsed does not imply anything about the quality or pace of reforms.

The rest of the report is organized as follows. Chapter 2 evaluates the electricity sector reforms' financial outcome and specifically their impact on the fiscal and quasi-fiscal operations of the government. Chapter 3 evaluates production efficiency, Chapter 4 the social outcomes, and Chapter 5 the environmental outcomes.

Chapter 2

Getting the Utilities off the Books

Reducing fiscal budget and quasi-fiscal deficits is a key measure to promote macroeconomic stabilization and growth in transition economies. Because the power sector is one of the largest contributors to these deficits in ECA, clarifying and then reducing government liabilities has been a driving force for sector reforms. The reforms can relieve budgetary pressure if they improve enterprise efficiency and reduce the need for transfers. They can also generate immediate budget revenues through asset sales during privatization.

This chapter analyzes the implications of reforms on the fiscal and quasi-fiscal balance of the government overtime. There are immediate budgetary impacts of reform (through privatization revenues) and companies or governments usually disclose this information (Annex B). Because the immediate budgetary impact of privatization revenues is most likely negligible in comparison with the long-term effect of reducing subsidies, it is not included in the analysis. In other regions the immediate fiscal impact of privatization has also been small. Pinheiro and Schneider (1994) show that in Latin America privatization did not result in substantial short-term fiscal benefits.

Quantifying the Sector Losses

There has been little systematic monitoring of the fiscal and quasi-fiscal balance of the power sector over the last decade, making it impossible to identify the magnitude of gains and losses. The scarcity of data may be due to the complexity of untangling financial flows between electric utilities and the government (Box 2.1). The lack of systematic monitoring is well recognized, and several new initiatives are under way to redress this situation, including a regional study that will include calculation of implicit subsidies in the power and gas sectors for ECA. This study will also estimate country specific cost-recovery tariffs.

This chapter quantifies the electricity sector's losses or implicit subsidy, using primary data and a formula developed by the World Bank (Box 2.2). It circumvents the problem of complex financial flows by defining the sector losses as the sum of three components: commercial losses,

> **BOX 2.1: UNTANGLING THE QUASI-FISCAL DEFICIT**
>
> The fiscal deficit is the difference between revenues and expenditures as recorded in the official government budget. In addition to fiscal deficit, public finance analysis takes into account government obligations that are not reflected in the budget but that result from explicit or implicit government liabilities outside the budget framework. Those could be explicitly recorded in legal documents or result implicitly from the logic of political events, from institutional rules, from social obligations of the government as understood by the public. These obligations, when they need to be financed, might cause substantial increases in government debt and budgetary spending due to required repayment of the debt. They thus constitute major fiscal (or budgetary) risks.
>
> When a utility is publicly owned, the government receives taxes and dividends from the utility and provides explicit and implicit subsidies, many of them not transparent. Untangling these financial flows requires detailed systematic data on financial flows that is not readily available. The data and analysis of the electricity sector fiscal and quasi-fiscal deficits are available for the countries for which the IMF or the World Bank undertook detailed studies, such as Armenia, Romania, and Russia (for example, see Petri and others 2002, Frienkman and others 2003, and Saavalainen and ten Berge 2003). There is also some evidence for Georgia and Moldova, but no systematic methodology or time series data have been available to date.

> **BOX 2.2: CALCULATING THE IMPLICIT SUBSIDIES (SECTOR LOSSES)**
>
> We use the "End-Product Approach" to calculating sector financial deficit, which is based on the quantity of energy sold, the end-user prices, and the collection rates compared with import or export prices. Developed by Infrastructure and Energy Group of ECA region, World Bank (ECSIE), the approach has been used in different modifications (such as Saavalainen and den Berge, 2003; Petri and others 2002; Cosse 2003). The modification used here is the latest one developed and applied by ECSIE in its current work.
>
> This approach allows calculation of the following three types of losses: commercial, collection, and tariff.[1] Commercial losses are the cost of electricity injected in the transmission system but not metered/billed minus the cost of electricity lost for technical reasons within nationally accepted norms for unavoidable losses. Measured this way, commercial losses include electricity stolen through illegal connections to the network and technical losses above norms due to bad maintenance and deteriorated physical assets. Collection losses are measured as the value of electricity billed but not collected from the consumers. Tariff losses are estimated as the difference between the amount billed (collected and not collected) to the consumers and the cost of the corresponding amount of electricity.
>
> The commercial and tariff losses depend on the assumptions about the cost recovery price (which in turn depends in large part on the value assigned to the sector assets). The higher the cost-recovery tariff, the greater the shortfall between this tariff and the actual tariff. Calculations in this study use the cost-recovery tariff of 4.8 US cents a kilowatt hour for all countries (Armenia, Azerbaijan, Georgia, Hungary and Poland) between 1992 and 2001, except for Kazakhstan, where the tariff is 4 cents until 1997 and 3 cents in the subsequent periods. This is an approximation of the long-term marginal cost, based on information provided by the local expert/investor estimates for this study. The cost-recovery tariff is estimated based on an allowance for generation, distribution, and the VAT. More precise estimates of cost recovery tariffs will be available on completion of a new regional study initiated by the World Bank.

collection losses, and tariff losses. The results are very sensitive to assumptions about the cost-recovery tariff, emphasizing the importance of the country-specific calculation of cost recovery tariffs if sector gains and losses are to be more accurately quantified.

The approach does not enable us to identify the shares of the sector losses accruing to the government and the private sector. This may be moot, however, because the government eventually ends up being responsible for financing most of these losses. For example, in

Georgia following privatization, the private sector share of these losses quickly fell while the government share grew. It may be that the private sector is transferring part of its deficit to the budget.

Deficits Declined for Different Reasons

Sector losses declined between 1993 and 2001 in all countries except Azerbaijan and Georgia, where it remained more or less unchanged after 1993 (Figure 2.1 and Annex C). By 2000 there

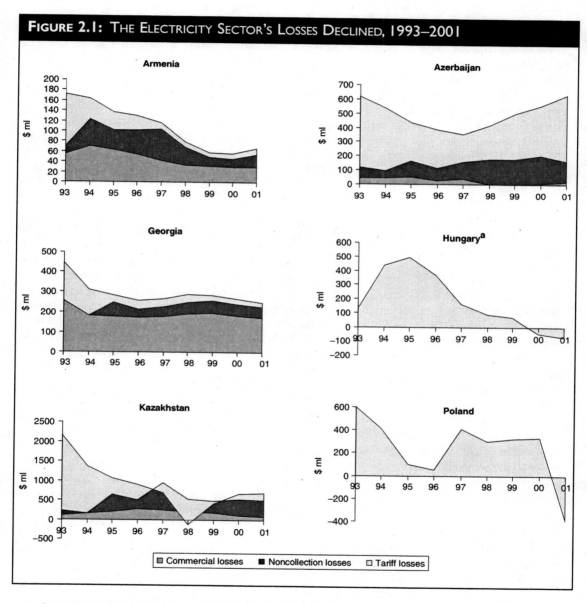

FIGURE 2.1: THE ELECTRICITY SECTOR'S LOSSES DECLINED, 1993–2001

a. According to the information provided by the local consultants for this study, most of the outstanding bill amount is collected within the first six months after the due date, and a negligible part of it—later, but within a year. Therefore, it is reasonable to consider the total bill amount as collected.

Note: Poland and Hungary have close to 100% collection rate

Source: Authors' calculations based on data provided by local consultants for this study.

was a slight surplus in Hungary. Sector losses were less than 1 percent of GDP in Poland and less than 5 percent in Armenia and Kazakhstan. It still remained high in Azerbaijan (10.6 percent in 2000) and Georgia (8.8 percent in 2000). (See Annex table 4 for details).

The decline is attributable to different reasons, suggesting that carefully targeted reform programs addressing specific deficit components work best. In Azerbaijan, targeting tariff losses by aligning actual tariffs with the cost-recovery price is key. In Georgia commercial losses are most important. In Armenia commercial losses and noncollection losses are equally important. In Kazakhstan noncollection losses are key.

Relationship Between the Electricity Sector Deficit and the Fiscal Deficit

The fiscal objectives of reforms are to reduce fiscal and quasi-fiscal subsidies to the electricity sector and to increase the transparency of the unavoidable subsidies by bringing them in the budget. Yet, the net fiscal gains of these changes may be small or negative, depending on the resources used to address the social implications of sharply rising electricity prices and reduced access to service. In addition, a tariff increase might result in increased non-payment of taxes, increased non-collection of billing, and reduced consumption, which can also hurt the budget. Moreover, the energy sector is fundamental to the economy. Energy reforms have serious implications for the production and profitability of most other sectors, affecting the flow of taxes to the budget.

For this reason the impact of reforms on the fiscal deficit is ambiguous, even though we observe a decline in the sector losses. Reform-induced transparency and accountability may reveal a truer picture of fiscal deficits in the short term, when the accumulated quasi-fiscal deficit of the sector emerges on the fiscal balance of the government. In the long term, we expect the reforms to increase the profitability of energy generation and distribution companies, and we expect the government to increase tax enforcement when these companies become privately owned. This means a higher flow of tax revenues from the energy sector to the budget. At the same time, electricity reforms reduce the profitability of the manufacturing subsector due to higher electricity tariffs, which means lower tax revenues for the budget from the manufacturing subsector. On balance, the budgetary impact will depend on the relative share of the energy and non-energy subsectors of the industry. The impact on the budget is likely to be positive in countries where a higher share of GDP is from the energy sector.

Consistent with the analysis in Chapter 1, the experience of the countries points to three adjustment paths. The first is the path of the high income energy importers (Hungary and Poland), the second, the oil exporting countries (Azerbaijan and Kazakhstan), and the third, the low income energy importing countries (Armenia and Georgia). Although at first glance the scatter plot of sector losses against country fiscal deficits appears to show a lack of correlation, a closer look reveals the difference in the country-specific adjustment paths (Figure 2.2). Hungary and Poland have low sector losses and fluctuating fiscal deficits. Armenia and Georgia show a decreasing tendency both in the sector losses and the fiscal deficits. The oil exporters Azerbaijan and Kazakhstan do not show any clear tendency, with the sector loss (sensitive to oil price) fluctuating widely.

The availability of domestic energy sources and the income of a country are crucial factors linking the sector loss with the fiscal deficit. With appropriate institutional conditions, in the countries with sufficient investment capital an increase in the price of electricity results in a higher level of investment in energy efficient production processes by the manufacturing sector (as in Hungary and Poland). If little investment capital is available for efficiency improvements, an energy importing country may be locked in a low efficiency-high sector deficit trap, when the investment capital is not available to improve efficiency (as in Armenia and Georgia). Energy rich countries (as in Azerbaijan and Kazakhstan) can use oil revenues to finance a high sector loss, because the energy sector is a major source of revenues for the budget.

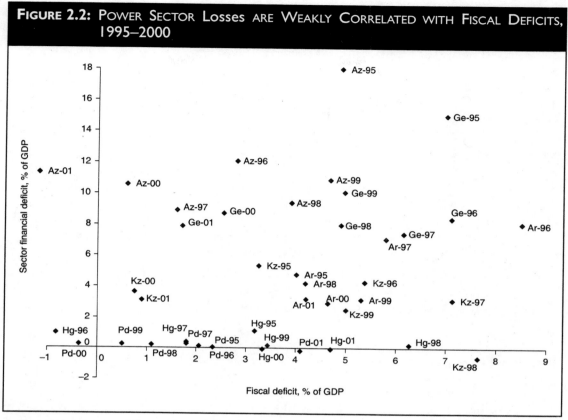

Note: Country names are abbreviated as follows – Armenia (Ar), Azerbaijan (Az), Georgia (Ge), Hungary (Hg), Poland (Pd), and Kazakhstan (Kz).

Source: Authors' calculations based on data provided by local consultants and on the public expenditure database for transition countries (World Bank 2002).

Reform Savings Did Not Go to Social Spending

For the government the net budgetary impact of reforms is equal to the savings from deficit reduction minus the costs of social assistance to mitigate welfare losses from higher electricity tariffs.[7] Unsurprisingly, countries with a high poverty incidence have very low per capita annual expenditures on social assistance (Figure 2.3).[8] If social assistance costs are lower than the budgetary gains from reducing the sector's losses, there appears to be room to increase the size of transfers to offset welfare losses from higher tariffs and for the donor community to plan an increasing role in facilitating social transfer mechanisms and providing technical assistance

---

7. This is a purely financial analysis of the budgetary impact. An economic analysis of the welfare impact on consumers would also need to take into account the gains to consumers from better service (if an improvement happens concurrently with a tariff increase).

8. Poverty is measured as the percentage of population below the absolute poverty line of US$2.15 (purchasing power parity) per capita per day. Poverty rates are from World Bank (2000b), p. 35. Reported absolute poverty rates are for 1999 in Armenia, Azerbaijan, and Georgia, 1998 in Poland, 1997 in Hungary, and 1996 in Kazakhstan. Social assistance/welfare expenditures are averages for 1997–2000 (calculated in US dollars).

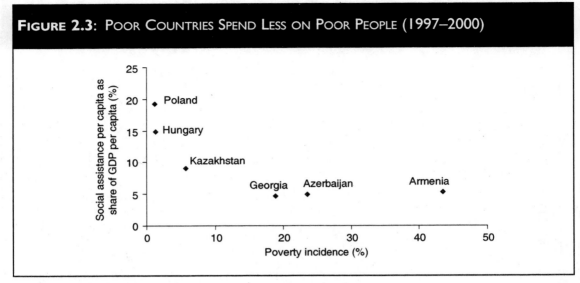

Note: Social assistance and GDP per capita are averages over 1997–2000.
Source: Public Expenditure Database for the transition countries, World Bank (2002).

to improve targeting. Of course this does not mean that transfers must be limited to electricity consumption, because it is possible for the public sector to make more productive investments in such other areas as education and health.

Estimating the costs of transfers is straightforward when social assistance is provided through special funds rather than through the general social assistance system. Special funds were created in Hungary and Georgia. In Hungary the justification was that the combined social and economic cost of disconnection would be higher than the cost of providing social assistance. In 1997 the Hungarian government contributed $4.6 million and the power companies $2.2 million to this fund. In Georgia the state provides subsidies through a complicated system of categorical privileges. The estimated annual cost of these subsidies at current tariff levels is $4.7 million a year, excluding administrative costs. In addition, in 1999 a program financed by the United States Agency for International Development (USAID) provided electricity in the winter to vulnerable households (World Bank 2003b, 2003c, 2003d). These costs are small compared with the financial sector deficit in Georgia and the surplus in Hungary.[9]

In Armenia, Poland, and Kazakhstan compensation for tariff increases is provided through the general social assistance program rather than earmarked funds. That makes identifying incremental social assistance costs a formidable task. They can rise because of an increase in social assistance payments per person that can be attributed to the reform or because of an increase in the total number of recipients qualifying for social assistance.[10]

The savings from a reduction in sector losses in ECA did not go to finance social protection or towards social expenditures on health and education, as it did in Latin America.

---

9. Clearly, these numbers are not directly comparable. Instead, the reduction in deficits should be compared with the costs of social assistance to mitigate the tariff increase. But the choice of a year would strongly influence the results. In Hungary the sector financial deficit fell by $205 million from 1996 to 1997 and by only $19 million from 1998 to 1999. In Georgia it fell by $5 million from 1998 to 1999 and by $17 million from 1999 to 2000.

10. This requires creating a counterfactual scenario of the poverty rate increase or reduction in the absence of the tariff increase.

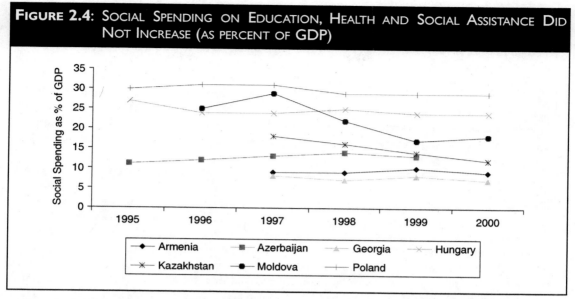

FIGURE 2.4: SOCIAL SPENDING ON EDUCATION, HEALTH AND SOCIAL ASSISTANCE DID NOT INCREASE (AS PERCENT OF GDP)

Note: Analysis of public expenditures in real terms reveals a similar trend, with falling or stable expenditure levels.
Source: Public Expenditure Database for the transition countries, World Bank (2002).

The explanation may lie in the different starting points of the countries in these two regions in the 1990s. The ECA countries accumulated enormous quasi-fiscal deficits in various sectors by the beginning of the transition period, and some of the savings from the reforms were probably spent to reduce these obligations. In Latin America fiscal pressure has been associated with an increase in the social spending (Box 2.3).

Conclusions

The legacy of central planning resulted in a power sector contributing to unsustainable government deficits. So getting the power utilities off the government books is the key to reducing fiscal budget deficits and promoting macroeconomic stabilization and growth in the transition.

Sector losses have been falling over the last decade, with tariffs the key determinant. Lessons include placing greater emphasis on ex-ante monitoring of sector deficits. Furthermore, using country-specific cost-recovery tariffs can generate more precise estimates of these losses. This work is currently under way in a separate regional initiative.

Despite the decline in sector losses, the impact of reforms on the fiscal deficit is ambiguous. This is due to a combination of factors, including the lack of data, very complex financial and budget flows in the sector, and the different adjustment paths of countries. Only a detailed analysis of the quasi-fiscal accounts would lead to a definitive conclusion about the impact of reductions in the sector loss on the fiscal deficit. On balance, we conclude that assessment of budgetary impact of the reform can be improved by carefully examining the relative share of the energy and non-energy subsectors of the industry. Positive outcomes are expected in countries where a higher share of GDP is from the energy sector. But institutional reform in these countries will also be the most difficult.

The savings from lower sector losses did not translate into an increase in social spending. The share of spending on social assistance, education, and health stayed the same or fell. It may be that savings from reforms went to reducing prior sector obligations. Aggregate social assistance data

> **BOX 2.3: SOCIAL SPENDING IN LATIN AMERICA INCREASED SINCE THE 1990S**
>
> In contrast to ECA, social spending in Latin America has increased 50 percent since the 1990s in the real terms, from the per capita spending level of an average of US$ 360 to $540 (CEPAL 2001). The highest increase took place in the countries with low and medium initial levels of per capita social expenditures. Interestingly, fiscal pressure has helped make public spending more socially oriented, increasing it as a share of GNP. Countries with high level of social spending in Latin America combine high fiscal pressure with a high priority for social spending, and vice versa. The fact that fiscal pressure led to budget cuts in the areas of public expenditure other than social spending shows that these countries pursued good policies. In ECA, fiscal pressure translates into a reduction in social spending both in real terms and in expenditure shares.
>
> Policymakers face a tradeoff between coverage and targeting of social assistance expenditures. An analysis of the composition of social expenditures in a sample of Latin American countries suggests that primary education, healthcare, and childcare programs are well targeted and inequality reducing. Pensions, housing, especially urban, and unemployment insurance mechanisms tend to be less pro-poor and less well targeted. Cash transfers, the most direct and non-distortionary way of providing assistance, require well developed administrative capacity and can be difficult to target. The disadvantage of other programs, such as the provision of infrastructure services and education, as a social assistance mechanism is that they are broad in coverage but not well targeted to the poor. Means testing may be a better way to target utility subsidies than lifeline tariffs. But, as suggested in studies reviewed by Estache (2001), the outcome depends on how well means-testing is implemented.
>
> To increase the efficiency of delivering social assistance, social protection must be carefully planned, with programs well targeted, well coordinated, and well designed to meet the specific needs of the population. To reduce the costs of means testing, Chile used the same targeting system for all types of programs, with applicants filling out one form that allowed them to qualify for a variety of programs. This approach proved very cost-effective, in 1996 a mere 1.2 percent of the benefits distributed.
>
> The inherent problem of social assistance and targeted spending is its highly cyclical nature. Protection falls during crises, when it is needed the most. Wodon and others (2003) estimate that a one percentage point decrease in GDP led to a reduction of two percentage point in targeted public spending per poor person in Latin America in the 1990s. It is therefore advisable for the governments to act in a long sighted way and to save at least some public funds during expansions in order to better protect the poor during recessions.

Source: Adapted from Wodon and others 2003.

show that these costs are much smaller than the savings generated by a reduced deficit. This implies that a larger share of savings can presumably go toward alleviating the blow from tariff increases. It does not imply that financing consumption (rather than more productive investments in health and education) is the best way to do this.

CHAPTER 3

CREATING MORE EFFICIENT COMPANIES

Electricity reforms are expected to improve the production efficiency of the power sector, ultimately translating into cost savings and service quality improvements for end users. They are also designed to make the sector financially sustainable by increasing the efficiency of resource allocation and the cost-effectiveness of sector investments. In theory the gains in efficiency materialize from a profit motive–created by more competition or transparent regulation. While the legacy of central planning suggests large potential production efficiency gains, the data collected here are ambiguous.

Improve Transparency and Accountability

Identifying the magnitude of production efficiency gains in ECA countries is confounded by changes in record keeping and accounting methods, by vested interests, and by private sector operators with few incentives to report production efficiency gains. Looking forward, Bank operations can improve transparency and accountability by emphasizing a systematic set of indicators in all sector operations and disseminating this information to the public. A best practice example is the Armenian Natural Monopoly Regulatory Commission, which discloses power sector performance indicators on the World Wide Web on a monthly basis.[11] And similar to the report cards used to elicit responses from citizens on public services in the Philippines and India (Bhatnagar 2001; Paul 1994, 1996), a system of citizen feedback on service delivery can be instituted. Such a mechanism can help create a more direct link between service quality and tariff increases. Another good example is the 2004 *World Development Report,* which outlines a system of accountability that connects consumers, government and providers through four interrelationships—client power, voice, compacts, and management.

The translation of improvements in production efficiency into cost savings and better service quality for end-users depends on assumptions about the counterfactual. One conclusion that can

---

11. Web address: http://rcnm.am. View indicators under "sector reports" link.

TABLE 3.1: SUGGESTED INDICATORS OF PRODUCTION EFFICIENCY

Stakeholder	Outcome Objective	Outcome Indicator	Examples
Consumers	Improved service quality	• Reduced number of outages	System average interruption frequency index[1]
		• Frequency and voltage stability	Number of deviations from established standards
Power sector (utilities)	Improved resource efficiency	• Increased revenue/collections	Rise in electricity billed as % of net supply; rise in collections as % of billings
		• Reduced cost of supply	Reduction in cost of generation ($/kilowatt hour)
		• Improved energy efficiency	Reduction on fuel use per kilowatt hour of electricity produced
		• Reduced losses	% reduction in (kilowatt hour lost/net kilowatt hour generated)
		• Improved operational efficiency	Rise in sales per employee, Rise in consumers served per employee
Government	Increased financial independence	• Increased sector investment (third party)	% increase in investment in generation/distribution/transmission
		• Reduced sector financial deficit	% decline in sector financial deficit expressed as a share of GDP

Source: Authors, based on review of project documents

be drawn immediately is that sector operations would benefit from simulations of realistic counterfactual scenarios. Assuming the service quality would have declined in the absence of reforms, the benefits that accrue to end-users are costs avoided. With tariffs well below cost recovery in most study countries, there is little to no expectation that efficiency gains would be passed on to end users through lower tariffs. This reinforces the need to place more emphasis on the design and use of outcome-based indicators of production efficiency as conditions in Bank and Fund operations. (Examples of outcome-based indicators are in Table 3.1.)

This chapter examines changes in six production efficiency indicators[12] over the last decade:

- Revenue per kilowatt hour (revenue/net generation).
- Cost of generation ($ per kilowatt hour generated).
- System losses [kilowatt hour lost/total kilowatt hour generated + (kilowatt hour imported—kilowatt hour exported)].
- Collections (revenues over billings).
- Operational efficiency (sales per employee, and consumers served per employee).
- Service quality (frequency, voltage, and average hours of service).

Revenue per kilowatt hour Rose

While profits cannot be determined from the available data, examining the cost of delivered electricity (revenue divided by net generation) provides a back-of-the-envelope estimate of profitability. Despite problems controlling losses, collections, and staffing, the revenue per kilowatt

---

12. Details of tariffs, capacity and efficiency indicators are presented in Annex E.

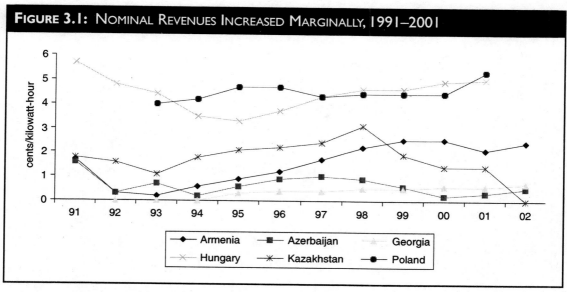

FIGURE 3.1: NOMINAL REVENUES INCREASED MARGINALLY, 1991–2001

Source: Authors' calculations based on data provided by local consultants.

TABLE 3.2: REPORTED GENERATION COSTS WENT DOWN, 1990–2002 (CENTS/KILOWATT HOUR)

	'90	'91	'92	'93	'94	'95	'96	'97	'98	'99	'00	'01	'02
Azerbaijan	3.0	1.1	2.8	3.0	0.6	1.9	2.2	2.5	2.3	1.9	1.8	1.8	1.5
Armenia	na	na	na	na	1.5	1.7	2.0	2.6	2.3	2.7	2.4	2.3	1.5
Georgia	na	na	na	na	na	na	na	na	1.1	1.3	1.2	1.0	0.9
Kazakhstan	0.7	na	na	1.4	2.2	2.6	3.2	na	na	2.5	2.2	1.7	na
Poland	na	1.9	1.9	1.9	2.3	2.6	2.7	2.6	2.9	3.0	2.9	3.1	na

Note: Generation costs are in nominal terms.
Source: Authors' calculations based on data provided by local consultants.

hour delivered (revenue/net generation) rose slightly from lows in the early to mid-1990s in almost all countries (Figure 3.1).

Generation Costs Declined

Reported generation costs (cents per kilowatt hour), one of the most important determinants of end-user tariffs, have exhibited a downward trend on average over the last decade. They have ranged from as low as 0.5 cents per kilowatt hour to as high as 4.3 depending on the country, year, and (most important) method used to calculate them (Table 3.2).[13] The analysis of

---

13. The economic costs of generation are often underestimated because initial investment expenditure is treated as a sunk cost and depreciation is not usually reflected in tariffs, the opportunity cost of fuel and electricity is disregarded, provisions for maintenance and capital repair are inadequate, fuel is often subsidized, and the reported cost of generation (especially in the early and mid-1990s) was often based on the cash accounting method, which records expenditures only when they are paid, an approach that can underestimate the real cost.

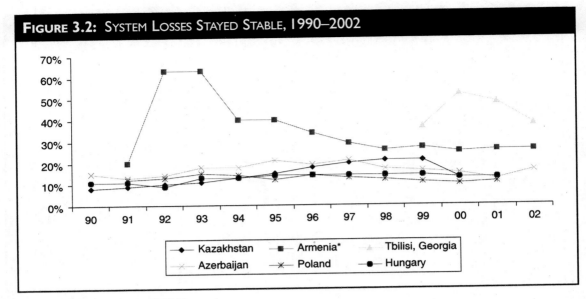

FIGURE 3.2: SYSTEM LOSSES STAYED STABLE, 1990–2002

* Expert estimates for 1992–93.
Source: Authors' calculations based on local consultant reports.

generation costs is complex: it is a function of international fuel prices, fuel and plan availability, dispatch impacts, and generation mix, factors largely beyond the control of the operator.

System Losses Held Steady

Technical and commercial losses (kilowatt hour lost/net generation) have held steady in the 10–20 percent range of net generation in four countries of the study, despite being the primary targets of donor conditions. The exceptions are Armenia and Georgia (Figure 3.2).[14] In Armenia losses fell by 40 percentage points from highs in the early 1990s. In Georgia total losses have not been quantified across the whole country, but anecdotal evidence suggests that commercial losses range between 50 and 60 percent in the regions. In Tbilisi reported losses since 2000 have ranged between 35 and 50 percent. Losses of 10–15 percent, as observed in Hungary and Poland, are consistent with fully commercialized electricity utilities. Continuing corruption and theft and the use of outdated distribution equipment keep losses above the desired levels in other countries. Theft continues to be a problem particularly because the low-voltage distribution systems that dominate much of the CIS make it simple to set up illegal connections to trunk lines.

Collection Rates Varied

Collection rates (revenues/billings), a function of both systemwide losses and tariffs, have varied widely. They have stayed close to 100 percent in Poland and Hungary, gone down in Azerbaijan, Kazakhstan, and most of Georgia (except in Tbilisi and Rustavi), and gone up in Armenia (Figure 3.3). Collection rates in Tbilisi have improved for distribution functions previously operated by AES Telasi. There have also been marked improvements in Rustavi under a pilot sponsored by USAID.

It is often postulated that private operators are more successful at improving collection rates than the public sector because they are not subject to the same political influences. While there is

---

14. According to the local consultants and other anecdotal evidence, actual losses in Azerbaijan and Kazakhstan are substantially higher.

> **BOX 3.2: OPPORTUNISTIC BEHAVIOR BY THE PRIVATE SECTOR?**
>
> "Almost by definition, certain features of regulated sectors make them more prone to renegotiation. First, regulation constrains the actions that a concessionaire can take, the most important being the setting of tariffs. Second, tariffs are expected to be set so that they allow the concessionaire to earn reasonable profits. When firms are not able to earn expected returns, it is rational for them to expect a change in contract terms. This is the premise behind the so-called financial equilibrium clause implicit or explicit in most concession service contracts and legislation. That is in principle a valid pillar of any concession contract, since private investors should be allowed to earn a fair rate of return on their investments. Having said that, it is also correct that the financial equilibrium ought to be subject to a number of provisos, including a conditioning to efficient operation. Yet the costs of providing services are rarely linked to a benchmark of efficient operations. And when they are, such costs are often disputed.
>
> "Strategic underbidding (or overbidding, depending on award criteria), to some extent encouraged by the incompleteness of contracts, also may explain the high proportion of renegotiation. As noted, many firms have won concession contracts by strategically underbidding (or overbidding), with the expectation that they would be able to renegotiate in the future, and governments have often been unable to commit to enforcing these agreements. If all potential bidders account for that possibility, an auction could still elicit the most efficient operator—but with significant underbids or overbids. But there are two problems with that argument. First, because renegotiation is a bilateral negotiation, the final outcome need not be guided by efficiency and welfare concerns, and rents could be transferred. Second, although any potential operator could submit a bid with the expectation of renegotiation, expectations might vary among bidders and not necessarily be correlated with their efficiency. Moreover, some enterprises may possess a systematic advantage in renegotiation and so be more likely to win a concession through underbidding (or overbidding). As a result a firm with high affiliation and high costs could win an auction."

Source: Guasch (2004).

Conclusions

Electricity reforms are expected to improve the production efficiency of the power sector, ultimately translating into cost savings and service quality improvements for end users. They are also designed to make the sector financially sustainable by increasing the efficiency of resource allocation and the cost effectiveness of sector investments. The efficiency gains materialize from a profit motive—created by more competition or transparent regulation.

While the legacy of central planning suggests large potential production efficiency gains, the data collected here on cost of generation, system losses collections, and operational efficiency are ambiguous. Despite continuing problems with losses, collection rates, and staffing, overall revenue per kilowatt hour has increased in almost all countries. The beneficiary of additional revenues from the growing gap between declining generation cost and rising retail tariffs is not so clear. Limited aggregate data suggests that service quality has improved in a number of capital cities.

Simple *ex-ante* simulations would allow better quantification of potential reform benefits and costs. And placing greater emphasis on outcome-based indicators of service quality will help ensure that future operations produce the intended end-user benefits. Both the private and the public sector can improve utility performance, suggesting that a broad range of contract arrangements can produce the desired outcomes. Finally, a number of local investors with both region and sector experience are filling investment needs, suggesting alternatives to the traditional definition of a strategic investor.

CHAPTER 4

MAINTAINING POWER TO THE POOR

ECA is unlike other regions of the world in that the socialist system gave almost all households access to reliable, inexpensive electricity. So, the welfare gains from increased access, one of the most immediate and tangible benefits of power sector reforms in other regions (Estache and others 2002; Galal and others 1994), is not a consideration in most ECA countries.

The electricity sector reforms have confronted households with a tradeoff between prices and service quality, and the perception of high political costs has had an adverse impact on the reform progress. Tariff increases and disconnections are very unpopular, and the public often views the sale of state assets to the private sector with skepticism. This is especially true when the public expects that the privatization of a public monopoly would result in a higher price without any improvement in the quality of service. If there is an improvement in service quality, it takes place a few years after privatization, while the price rises early on in the process. This mismatch between the timing of the gains and costs of reform and the uncertainty about the gains causes consumers to be skeptical. Experience in Latin America shows that privatization can succeed when the social issues are well addressed—and fail when combined with inadequate analysis of social and distributional issues and a lack of political will to support the reform.[17]

Even though the official statistics and household surveys suggest that access to service is nearly universal, supply is often rationed. The low investment in electricity generation and

---

17. Some anecdotal evidence from Latin America and the ECA countries illustrates this. One example is a failed privatization of the water and sanitation services of Tucuman in Argentina in 1995. The concessionaire bid an immediate 68 percent tariff increase to fund the required investment program. The tariff increase was to affect all consumer groups equally. In response, consumers organized a nonpayment campaign, forcing the investor out of business (Estache and others 2001). Another example is the failed privatization of the Almaty electricity distribution company in Kazakhstan in 1996. The contract specified a tariff increase in return for investment requirements by the concessionaire. Facing strong opposition to the tariff increase, the government failed to fulfill its contractual obligations, and in response the concessionaire did not make the specified investments and walked away after arbitration in 2000 (news sources).

TABLE 4.1: URBAN NETWORK ENERGY USE IN ECA (PERCENT OF HOUSEHOLDS)

Country	Central Heat		Central Gas		Electricity	
	Poor	Nonpoor	Poor	Nonpoor	Poor	Nonpoor
Armenia (1999)	11	14	4	16	97	99
Croatia (1997)	15	39	19	30	99	100
Kyrgyz Rep. (1999)	17	55	13	33	100	99
Latvia (1997)	70	83	57	68	99	100
Lithuania (1998)	31	46	47	56	85	94
Moldova (1999)	17	57	37	70	65	89
Tajikistan (1999)	1	1	3	6	100	100

Source: Lampietti and Meyer (2002).

distribution infrastructure in the last decade resulted in severe deteriorations in service quality. Some countries experience frequent interruptions in electricity supply. Others experience voltage fluctuations that destroy household appliances (Markandya and others 2001). Unless investments are made in rehabilitation and maintenance of the infrastructure, households may experience widespread supply shortages in the future.[18] This also leads us to our first conclusion that greater emphasis must be placed on explicitly linking tariff increases with improved service quality in order to minimize negative welfare effects. This is also likely to generate more political will to support the reform.

Household Energy Use Patterns Differ

There are systematic differences in energy use in urban areas in the ECA region. Separating network and non-network energy use provides insight into energy use patterns (Table 4.1). Almost all households use electricity, with small differences between the poor and the nonpoor. But poor people use much less central heat and gas. Are the nonpoor more likely to use network energy because they have better access to the network, or is it because they make different choices? Although this question cannot be answered with the data from household surveys, it points to the need for country-specific analysis to identify the supply constraints—such as network location and capital equipment (such as gas heaters)—that limit poor people's access to network energy.

If poor people are not using network energy, what are they using? Primarily dirty non-network energy (Table 4.2). Wood and coal use are consistently higher among the poor—except in Tajikistan, where coal is heavily subsidized for everyone. Except in Latvia, the nonpoor are more likely to use liquefied petroleum gas (LPG), the cleanest non-network energy. The poor may favor dirty non-network energy because it is less expensive or because they do not have the resources to spend on network appliances. Yet as noted, burning dirty fuels has social costs—mainly air pollution and deforestation—that require careful, country-specific analysis of the economic implications of raising the price of clean energy (Lampietti and Meyer, 2002).

---

18. Cambridge Energy Research Associates estimates that half of Russia's generation capacity must be retired in the next 20 years as it reaches the end of its productive life, while more than the total installed generation capacity of France needs to be added. If these investments are not made, Russia is expected to suffer from nationwide electricity shortages in the near future. ("Russian Electricity: In Need of Shock Therapy." *The Economist*. August 31, 2002. pp. 50–51.)

TABLE 4.2: URBAN NON-NETWORK ENERGY USE IN ECA (PERCENT OF HOUSEHOLDS)

Country	LPG		Kerosene		Coal		Wood	
	Poor	Nonpoor	Poor	Nonpoor	Poor	Nonpoor	Poor	Nonpoor
Armenia (1999)	17	27	14	11	n/a	n/a	47	50
Croatia (1997)	44	45	3	7	1	1	51	26
Kyrgyz Rep. (1999)	24	39	31	17	60	31	46	22
Latvia (1997)	37	28	n/a	n/a	<1	<1	1	2
Lithuania (1998)	n/a	n/a	n/a	n/a	<1	<1	1	2
Moldova (1999)	6	7	n/a	n/a	9	5	12	9
Tajikistan (1999)	n/a	n/a	<1	1	11	18	47	32

Note: n/a—not available from household survey
Source: Lampietti and Meyer (2002).

Tariffs Rose

In an effort to reach cost-recovery levels, the residential electricity tariffs were increased in all countries[19] In nominal dollar terms they nearly doubled in Azerbaijan, Georgia, and Hungary between the mid-1990s and 2002; Poland and Hungary currently have the highest tariffs. In Kazakhstan, Azerbaijan, and Moldova the tariffs first increased or remained stable and then fell in the late 1990s due to the depreciation of local currency and the 1999 devaluation triggered by the financial crisis in Russia. In real terms the tariff increase during this period was the highest in Azerbaijan and Georgia, where the tariffs quadrupled and tripled, respectively. By 2002 the tariffs doubled in Armenia and Moldova, increased by about 160 percent in Hungary, remained unchanged in Poland and fell in Kazakhstan (Figure 4.1). Such substantial tariff increases are unlikely to be welfare neutral unless accompanied by improvements in service quality or cushioned by income transfers.

A complete analysis of the tradeoffs between tariffs and service quality requires comparing the welfare losses from price increases with the gains from improved service quality (which may take place with a lag). As noted in chapter 3, aggregate service quality is thought to have improved since the early 1990s, the crisis period in Armenia, Azerbaijan, and Georgia, implying welfare gains for consumers.[20] In Kazakhstan service quality remained unchanged. But it is not clear what the situation would have been in the absence of privatization of generation facilities, which were in dire need of investments. It is also not clear how much of the improvement, where it did take place, can be attributed to the general political and economic changes, and how much to power sector reforms. So instead of evaluating the welfare gains from service quality improvements, the welfare impact of reforms is first assessed by examining changes in the share of electricity expenditures in income (using total expenditure as a proxy), particularly for the poor and nonpoor and then by assessing the loss in consumer surplus.

---

19. In Kazakhstan tariffs also increased in nominal terms after 1993, but the increase was negligible, unlike the other countries in this study.

20. In Georgia supply improved in Tbilisi after privatization of the distribution company. The company was not always able to provide 24-hour supply for reasons beyond its control, such as interconnectedness of the power grid with the rest of the country. Power supply in other areas, where distribution is still publicly owned, is worse. (Project Appraisal Document for an Electricity Market Support Project in Georgia, World Bank, April, 2001. p. 5)

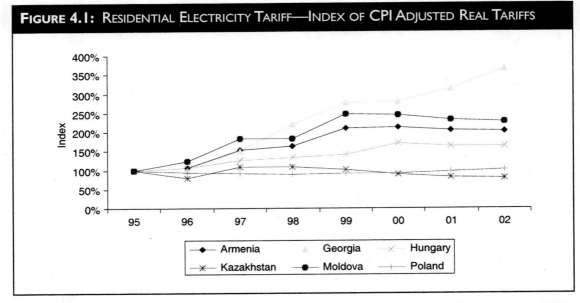

FIGURE 4.1: RESIDENTIAL ELECTRICITY TARIFF—INDEX OF CPI ADJUSTED REAL TARIFFS

Note: Nominal tariffs in local currency were adjusted by the CPI. Base year is 1995 (1996 for Georgia). 1995 is the first year for which data for all countries are available. In Poland and Hungary most of the tariff increase occurred before that. Azerbaijan is not included in the graph. Though its nominal tariffs, when calculated in US cents/kWh, is among the lowest in the regions, the CPI adjusted real tariff is the highest.

Source: Authors' calculations based on data provided by local consultants and Counterpart International (for Moldova).

The Burden Increased

Mean expenditures on electricity range from 2 to 10 percent of total expenditure on average, with a high of 14 percent for the bottom quintile in Armenia in 2001 and a low of less than 1 percent for the bottom quintile in Kazakhstan in 2001 (Table 4.3). The share increased slightly in all countries and for all welfare quintiles.[21] Bottom quintile annual change in electricity share varied between −1.9 percentage points (Poland, 2000) to 41 percentage points (Moldova, 1999). The results are conditional on the households reporting positive expenditures on electricity.[22] So, the quality of electricity expenditure data collected in the household budget surveys and living standards measurement studies affects the ability to draw generalizable conclusions.

The increasing share of expenditures on electricity can be explained by rising tariffs, falling income, reduced rationing (as service quality improved), or inelastic demand. That the share falls monotonically across the welfare distribution is consistent with findings that the poor devote a

---

21. Average yearly change in electricity shares was positive for bottom, top, and all quintiles for Georgia (1997–2002), and Poland (1994–2001). For Moldova, it declines from 2000. There was a similar positive change in electricity shares for all quintiles in Armenia (between 1999 and 2001), Hungary (between 1993 and 1998, 1998 and 1999), and Kazakhstan (between 1996 and 2001).

22. The share of households with zero expenditures varies across surveys due to different survey design. Surveys in which respondents are asked about the previous month's expenditures frequently have over half the respondents reporting zero expenditures, a result more likely due to the nature of the question than to the fact that none of them pay for electricity. If the whole sample is included, it is not clear whether the calculated shares of electricity expenditures in income vary due to different survey structures or because the share actually changed. So only the subsample with households with positive expenditures is included. A series of tests reveals that these households are systematically different from the other households in the sample. Household size and total expenditure levels are similar for both groups, while the share of urban households is substantially higher in the group with non-positive expenditures (excluded from the calculations). The sample thus underrepresents urban households.

TABLE 4.3: SHARES OF SPENDING ON ELECTRICITY WENT UP, 1993–2002
(MEAN OF HOUSEHOLD ELECTRICITY SHARES)

	1993	1994	1995	1996	1997	1998	1999	2000	2001	2002
Armenia[a]										
Bottom quintile	na	na	na	18.2	na	na	10.2	na	13.9	na
Top quintile	na	na	na	3.6	na	na	5.3	na	6.5	na
Mean	na	na	na	9.0	na	na	7.2	na	8.8	na
Georgia										
Bottom quintile	na	na	na	3.1	3.8	5.7	4.4	3.7	3.6	6.3
Top quintile	na	na	na	1.5	1.5	1.6	1.5	1.5	2.3	2.0
Mean	na	na	na	2.0	2.3	2.9	2.5	2.4	2.7	3.4
Hungary										
Bottom quintile	4.3	na	na	na	na	5.6	6.3	6.5	na	na
Top quintile	2.2	na	na	na	na	3.6	3.7	3.7	na	na
Mean	3.4	na	na	na	na	4.9	5.4	5.4	na	na
Kazakhstan										
Bottom quintile	na	na	na	0.16	na	na	na	na	0.92	na
Top quintile	na	na	na	0.11	na	na	na	na	0.56	na
Mean	na	na	na	0.14	na	na	na	na	0.69	na
Moldova										
Bottom quintile	na	na	na	na	6.5	7.6	10.8	8.4	7.7	6.3
Top quintile	na	na	na	na	2.7	3.2	4.2	4.1	3.7	3.0
Mean	na	na	na	na	4.1	4.6	6.1	5.5	5.2	4.3
Poland										
Bottom quintile	4.5	4.6	4.6	4.5	4.5	6.6	7.2	5.3	5.8	na
Top quintile	2.4	2.3	2.4	2.3	2.1	3.5	3.7	2.6	2.9	na
Mean	3.4	3.4	3.4	3.3	3.3	3.8	4.1	4.1	4.3	na

a. For 1999 the results reported are based on 1999 Household Budget Survey. The results for the 1999 Energy Survey for Armenia are similar: the average expenditure share is 6.2 percent, that of the bottom quintile 8.8 percent, and that of the top quintile 5.0 percent.

Source: Authors' calculations from household survey data.

higher share of total expenditures to energy (Lampietti and Meyer 2002) and that electricity is a necessary good. It also implies a greater proportionate welfare loss for the poor and a more active search for substitutes when electricity tariffs increase.

The overall welfare impact of the reforms can be measured by the change in consumer surplus. Consumers gain from an improvement in service quality and the removal of rationing, but lose from an increase in price. Since we cannot measure the gains from the service quality improvement from the existing data, we focus on the consumer surplus change of a price increase. The magnitude of the welfare effect of a price increase depends on the household's dependence on the energy source, measured by its budget share, the price change, and the household's access to substitute energy sources and other goods and services. The third is measured by the elasticity of demand. For electricity the elasticity of demand is typically low: it has been estimated between –0.08 and –0.32 for a range of countries (Hope and Singh, 1995).[23] The

---

23. Electricity price increase of the late 1980s for Zimbabwe, Colombia, and Turkey.

TABLE 4.4: Consumer Surplus Fell

Country	Real Price Increase		Starting Point Budget Share	Lost (gained) Consumer Surplus (Percent of Total Household Budget)		
	Period	% Change	%	$\eta = -0.1$	$\eta = -0.5$	$\eta = -0.9$
Armenia	1996–2002	98%	9.0	−8.38	−8.39	−8.40
Georgia	1996–2002	267%	2.0	−5.33	−5.33	−5.33
Hungary	1993–2002	57%	3.4	−2.36	−2.36	−2.36
Kazakhstan	1996–2002	2%	0.2	0.00	0.00	0.00
Moldova	1997–2002	43%	4.1	−0.17	−0.17	−0.17
Poland	1994–2002	9%	3.4	−0.14	−0.14	−0.14

Source: Budget shares calculated from household survey data.

budget share of electricity in household income varies depending on income and the geographic location of a household. Typically network energy budget shares are inversely related to income, with the poor urban households spending a highest share of their total income compared to all other groups of households (Hope and Singh 1995). This is also the case in these countries in ECA.

We present the consumer surplus calculation[24] in a range of demand elasticity scenarios. Zero elasticity means that the household is unable to adjust, so this provides an upper bound of the welfare loss. At a price elasticity of −1, the household is able to reduce electricity consumption in response to the price increase, so reducing the magnitude of the welfare loss. Households in Armenia and Georgia experienced the largest welfare loss, expressed as a percentage of total household budget, because of the high magnitude of the price increase and the high shares of electricity expenditures in the total budget (Table 4.4). In real terms the electricity price fell slightly in Kazakhstan, so households did not experience a welfare loss.

Consumption is Low

Studying consumption using household survey data is confounded by the presence of arrears (nonpayments), which make it impossible to determine whether reported electricity expenditures represent current or historical consumption. But it is possible to examine a sample of household electricity consumption records from distribution companies in Armenia and Georgia.[25] These data do not take into account households not included in the utility's database and thus might be an underestimate of household consumption. In Armenia, where service quality has been consistently high since 1996 (provided 24 hours a day, seven days a week), mean household consumption fell steadily from an average of 160 kilowatt hours (kWh) a month in 1998 to 117 kWh a month in 2001.[26] In Tbilisi, where service quality has been improving since 2000, consumption remained constant at about 150 kilowatt hour a month (Figure 4.2). Greater seasonal fluctuations

---

24. In most cases, the welfare loss or gain calculation includes information on access. For the ECA region, access to electricity is close to 100 percent. The welfare loss calculation should also include information on disconnection as a result of price rise, but we could not get accurate data.

25. In Armenia, the sample consists of (usable) monthly records for 1,197 households for the period from March 1998 to September 2002, and in Georgia for 288–408 households, depending on the month, for the period from January 2000 to September 2002. Data were provided by electricity utilities.

26. The median consumption for the same period fell from an average of 143 kWh a month to 95 kWh a month.

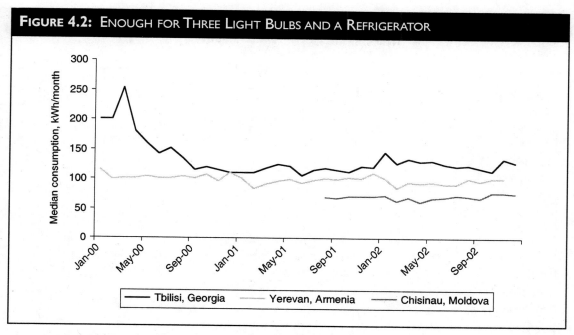

FIGURE 4.2: ENOUGH FOR THREE LIGHT BULBS AND A REFRIGERATOR

Note: Calculations include only households with positive electricity consumption.

Source: Authors' calculations using AES Telasi data (reported in Lampietti and others (2003)) and electric utility records for 1,197 households in Yerevan and 2,092 households in Chisinau.

in electricity consumption in Armenia than Georgia suggest greater access to inexpensive substitutes for heating in Georgia.

Household electricity consumption is close to basic minimum needs, sufficient only for lighting and refrigeration.[27] The median consumption was as low as 84–100 kilowatt hour a month during 2002 in Armenia and Georgia. The assumption is that an electricity demand function is kinked, as is characteristic of necessary goods, sloping steeply around the minimum required for basic needs and then rapidly leveling off as the quantity of electricity consumed moves from necessity to luxury.

This leads to two conclusions. First, the welfare losses from a price increase are high for households with very low electricity consumption. Second, there may now be little scope for efficiency gains from the household sector in Armenia, Georgia, and Moldova. There was such scope at the beginning of the transition because the former CIS countries have traditionally been energy intensive, but the move toward cost-recovery tariffs and regulatory improvements had little scope for further reductions in consumption. The decline in household incomes associated with transition also added to the decline in electricity use. If the price of electricity increases further, there will only be a small reduction in household energy consumption.

Gas May Be Filling the Gap

Electricity is the single largest item in the household energy budget in Armenia, Georgia, Moldova, and Poland, and the share is largest for the bottom quintile (Table 4.5). Assuming that it consumes less than the top quintile, this implies that electricity is used for the most basic needs for which other fuels are poor substitutes, so demand by the poor is inelastic.

---

27. A refrigerator (manual defrost, 5–15 years old) consumes about 95 kWh a month and 3 incandescent light bulbs another 30 kWh a month.

TABLE 4.5: ELECTRICITY EXPENDITURE AS A SHARE OF TOTAL ENERGY EXPENDITURE, 1993–2002 (MEAN OF HOUSEHOLD ELECTRICITY SHARES)

	1993	1994	1995	1996	1997	1998	1999	2000	2001	2002
Armenia[a]										
Bottom quintile	na	na	na	na	na	na	73	na	96	na
Top quintile	na	na	na	na	na	na	57	na	88	na
Mean	na	na	na	na	na	na	62	na	90	na
Georgia										
Bottom quintile	na	na	na	83	65	68	68	68	67	na
Top quintile	na	na	na	54	38	41	40	40	45	na
Mean	na	na	na	65	48	51	53	53	53	na
Hungary										
Bottom quintile	32	na	na	na	na	37	38	39	na	na
Top quintile	33	na	na	na	na	37	38	39	na	na
Mean	31	na	na	na	na	36	37	38	na	na
Kazakhstan										
Bottom quintile	na	na	na	41	na	na	na	na	47	na
Top quintile	na	na	na	30	na	na	na	na	65	na
Mean	na	na	na	34	na	na	na	na	56	na
Moldova[b]										
Bottom quintile	na	na	na	na	86	87	92	89	87	86
Top quintile	na	na	na	na	80	74	77	81	79	76
Mean	na	na	na	na	83	80	84	86	84	81
Poland										
Bottom quintile	62	61	61	60	60	68	68	67	69	na
Top quintile	35	33	33	32	32	39	41	43	44	na
Mean	47	45	45	43	44	39	40	43	41	na

a. The 2001 survey, unlike the 1999 survey, does not capture well energy sources other than electricity.
b. The share is high in Moldova because none of the surveys captures a larger portion of other energy sources.
Source: Authors' calculations from household survey data.

Given the large price increases and the inelastic demand, it is surprising that expenditures went up only 1.5 percentage points on average. The impact may have been mitigated by improvements in service quality or substitutions of other energy sources. While there are no perfect substitutes for electric lighting, refrigeration, and television, given a choice of substitutes for electricity in heating and cooking, households are likely to choose natural gas because it is clean, convenient, and the low priced. Other alternatives—such as kerosene, coal, and wood—are less convenient.

Relative fuel prices and fuel availability influence household energy consumption choices. Even at full import prices gas is substantially less expensive than electricity (Figure 4.3). While there may be additional costs associated with the technology required to use gas (metering and gas-fired appliances), the convenience and savings suggest that, given access, it is the household fuel of choice.

Back-of-the-envelope calculations confirm the rising use of natural gas. In Armenia residential consumption of natural gas more than tripled from 1996 to 2001 (from 29,000 tons of oil equivalent to 90,000), while monthly electricity consumption dropped from 187,000 tons of oil

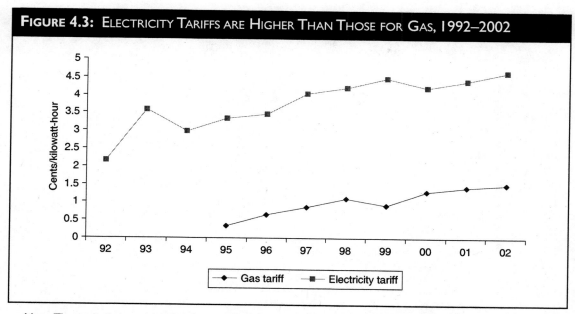

FIGURE 4.3: ELECTRICITY TARIFFS ARE HIGHER THAN THOSE FOR GAS, 1992–2002

Note: The applied conversion factor was 277.8 kWh per Giga-Joule (GJ) of natural gas (International Energy Agency). Average tariffs were calculated for Armenia, Azerbaijan, Georgia, Hungary, Kazakhstan, Moldova, and Poland. Note that this is a simple average. The number of observations varies by year depending on data availability.

Source: Authors' calculations based on data from local consultants, Counterpart International (for Moldova), and ERRANET database

equivalent to 106,000.[28] For Georgia the number of gas connections in the capital quadrupled from 2000 to 2003.[29]

Given the potential welfare losses of rising electricity tariffs households cooking and heating with electricity should be encouraged to use clean and inexpensive substitutes, such as natural gas. This can be done through a variety of instruments, as long as the government explicitly compensates the utility for any social transfers it provides. An example would be bidding out competitive subsidies to encourage the extension of natural gas networks to poor neighborhoods.

Enforcement is Necessary

Nonpayments, or arrears, are one of the most vexing problems. A key reform objective has been to resolve it, and collection rates indeed appear to rise after privatization (Box 3.1).[30] Understanding who accumulates arrears has important implications for the welfare effect of reforms. If it is mainly the poor, affordability may be a problem and special care must be taken by the state to provide adequate assistance to the poor. If it is all households, free-riding may be the problem and stricter enforcement will not disproportionately hurt the poor.

---

28. Total residential consumption from the energy balance data in Armenia (Ministry of Energy). Converted to kilowatt hours using the conversion factor of 1,000 kWh = 0.086 tons oil equivalent, this is equivalent to an increase in natural gas consumption from 337 million kWh in 1996 to 1,046 million in 2001, and a reduction in electricity consumption from 2,174 million kWh in 1996 to 1,232 million in 2001 (conversion factor is from the World Energy Council).

29. Tbilgazi's customer base increased from 39,000 households in June 2000 to 164,000 households in January of 2003 (Lampietti and others 2003).

30. Lampietti and others (2003) reach a similar finding, using collection rates (figure 13, p. 30).

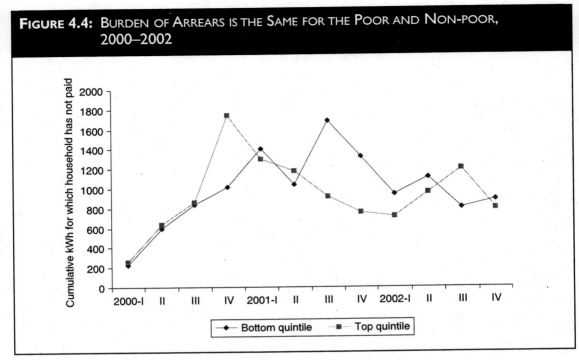

Figure 4.4: Burden of Arrears is the Same for the Poor and Non-poor, 2000–2002

Source: Authors' calculations based on AES data.

Evidence on the absolute amount of cumulative arrears (in kilowatt hours) in Georgia suggests that the problem may be free-riding. In some periods cumulative arrears[31] of the bottom quintile exceed those of the top quintile,[32] but in others there is no statistical difference between the two groups (Figure 4.4). However, the cumulative arrears of the bottom quintile are higher than those of the top when arrears are measured relative to electricity consumption. The top quintile consumes more than the bottom quintile in 6 of 12 quarters (all quarters of 2002), and this difference is statistically significant.[33]

Searching for Better Transfers

In addition to improving service quality, the government can mitigate the welfare effects of price increases by providing assistance to vulnerable households and by stimulating income growth. There is much debate about the validity of each assistance measure. Lovei and others (2000) found that instruments that perform well on some criteria tend to perform poorly on others. Furthermore, not all subsidy mechanisms are applicable or perform equally well across all countries and utility services. So no single instrument has been identified that would outperforms all others.

One of the most contentious debates is between lifeline tariffs, which subsidize an initial block of electricity for all users, and direct income transfers. Proponents of direct transfers argue that lifelines are not targeted and thus encourage inefficient energy use. Opponents claim that

---

31. Cumulative arrears take into account that some households may carry arrears for many months before being disconnected.
32 According to t-test results, this difference is significant in the fourth quarter of 2000 and in the third and fourth quarters of 2001. In the fourth quarter of 2000 arrears of the top quintile exceed those of the bottom, and in 2001 the reverse is true.
33. According to t-test results, this difference is significant in the fourth quarter of 2000, the fourth quarter of 2001, and in all quarters of 2002.

Box 4.1: Simulation of Alternative Subsidy

Energy subsidies to Georgian households are available through a range of programs. One provides all veterans and pensioners between 35 and 70 kWh per household a month (recently increased to 240 kWh a month in the winter and 120 kWh a month in the summer). Refugees and internally displaced persons (including those not living in collective centers) also receive substantial quantities of free electricity. Other government programs provide households 850 m³ of natural gas a year. In addition to the government-funded subsidies, a major donor-financed subsidy to electricity customers—Winter Heat Assistance Program (WHAP)—has been in place for the past five years. WHAP finances the supply of electricity to low-income households in Tbilisi for winter heating during the January–April period. The amount each household receives has varied each year depending on the funding available. It was 850 kWh in 2000 and 1000 kWh in both 2001 and 2002.

The proposed subsidy would reach a higher percentage of low income households than either of the existing subsidies (Table 4.6). It would also reach a higher percentage of the other quintiles as well. The absolute subsidy to each household would be substantially lower than in either of the existing programs. The total cost would fall between the WHAP and the government program. The new program would thus be more cost-effective (in Georgian Lari per household) than either of the existing programs.

The simulation illustrates an alternative subsidy design, but there are several important caveats. First, the cost of the proposed subsidy would increase as the old subsidy is phased out, reducing some of the fiscal benefits. This is because more households are likely to consume in the 75 to 125 kWh range. At the same time, poverty targeting may well improve as the old subsidy is phased out. With the loss of the existing subsidies, consumption will be based more directly on actual household income. Second, several well organized stakeholders are encouraging the government to keep the subsidies in place, including veterans (who do not wish to lose their benefits) and Telasi (which presumably enjoys the simplicity and predictability of payments associated with the current system). Third, these results are based on data from Tbilisi, and caution must be taken in generalizing them to the rest of the population.

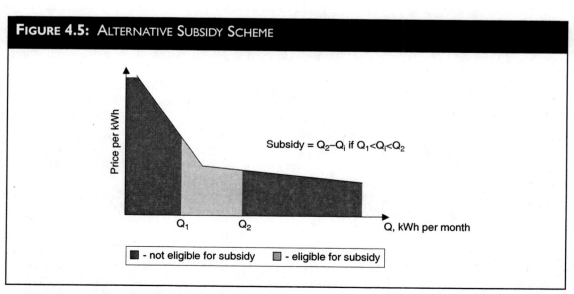

Figure 4.5: Alternative Subsidy Scheme

Subsidy = $Q_2 - Q_i$ if $Q_1 < Q_i < Q_2$

Note: Q_i = household consumption in period i
Source: Authors.

TABLE 4.6: SIMULATION OF SUBSIDY COST-EFFECTIVENESS FOR TBILISI, GEORGIA

	Bottom	Mid-Low	Quintile Middle	Mid-High	Top
Households receiving (percent)					
Government subsidy	10	16	18	11	10
WHAP subsidy	27	23	21	19	14
Proposed subsidy[a]	44	38	40	42	39
Proposed subsidy—no gas users[b]	40	35	43	34	35
Average subsidy per household (kWh/yr)					
Government subsidy	610	561	548	646	535
WHAP subsidy	1000	1000	1000	1000	1000
Proposed subsidy[a]	407	411	497	476	324
Proposed subsidy—no gas users[b]	398	384	479	382	287
Cost effectiveness (GEL/household)					
Government subsidy	76	70	68	80	66
WHAP subsidy	124	124	124	124	124
Proposed subsidy[a]	50	51	62	59	40
Proposed subsidy—no gas users[b]	49	48	59	47	36

a. The proposed subsidy is for households that consume between 875 and 1750 kWh a year. These households are given a monthly subsidy equal to the difference between 125 kWh and their monthly consumption. The assumed tariff is 0.124 Georgian Lari/kWh.

b. The second proposed subsidy is identical to that described in (a), except that it is available only for households that do not have access to natural gas.

Source: Lampietti and others 2003.

transfers through the general social assistance system, while theoretically attractive, fail to reach a large share of the poor because of inadequate targeting.

Income transfers tend to be well targeted in countries with less than 10 percent of the population below the poverty line, with enough funds to finance the administration of social assistance, and with a small informal sector so that means testing is easy. Examples include Hungary and Poland.[34] The transfers are less well targeted in countries such as Armenia and Georgia, where nearly half the population is poor, budget resources are insufficient, and means testing is very difficult. The case for lifeline tariffs is stronger in countries with high poverty rates and poor targeting—so long as there is sufficient political will to keep the size of the blocks small (say below 50 or 100 kWh) and to reimburse the utility for its costs—than in the countries with low poverty rates and well targeted social assistance.

Where tariff-based subsidies are in use, it may be possible to reorient their design to maximize consumer welfare gains and minimize the cost to the government budget. An electricity demand function is kinked, sloping steeply around the minimum required for basic needs and then rapidly leveling off as the quantity consumed moves from necessity to luxury. Ideally a subsidy would provide just enough compensation to ensure that each household consuming in the

---

34. With a poverty line of $2.15 a day, the poverty headcount index is 44 percent in Armenia, 24 percent in Azerbaijan, 19 percent in Georgia, 6 percent in Kazakhstan, and less than 2 percent in Hungary and Poland. With a poverty line of $4.30 a day, these numbers change to more than 50 percent in Armenia, Azerbaijan and Georgia; 30 percent in Kazakhstan—and less than 20 percent in Poland and Hungary (results of household surveys, reported in World Bank [2000a]).

steeply sloping (inelastic) portion of the demand curve (where welfare losses are large) consumes in the flat (elastic) portion where welfare losses are small. Since the exact location of the kink[35] cannot be identified, an upper (Q_2) and lower bound (Q_1) approach can be used (Figure 4.5). The lower bound eliminates incentives for gaming the system (such as installing multiple meters) and excludes residences with very low electricity usage, such as houses that are not primary residences. The utility would implement the program, using their billing database, and the government would be responsible for transferring the necessary resources to them. A simulation in Georgia indicates that such a program would be more cost-effective than existing categorical and income-targeted electricity transfers (Box 4.1).

Conclusions

The socialist system gave almost all households access to reliable, inexpensive electricity. So, welfare gains from increased access—one of the most immediate and tangible benefits of power sector reforms—is not a consideration in most ECA countries. The welfare gains come from improvements in service quality, again reinforcing the need for reform to emphasize outcome-based indicators of service quality.

Without disaggregated baseline data on service quality, reform appears closely linked to a fall in welfare. Electricity spending as a share of income increased, especially for the poor, while consumption stayed the same. Several lessons emerge. In many cases, tariff increases can and should be explicitly timed to coincide with service quality improvements. Yet, this may not always possible. Where it is not, the adverse impact of tariff increases, especially on low-income consumers, can be mitigated by improving access to and efficiency in the use of clean alternatives. In some locations, especially urban areas where households heat with electricity, natural gas may be a viable substitute.

There are serious problems associated with self-reported electricity data (as well as energy expenditure data in general) collected in the traditional poverty monitoring surveys such as the Living Standard Measurement Studies and Household Budget Survey. The reason is that the questions are confounded by recall error, under and over reporting, and the presence of arrears, making it impossible to identify current and historical consumption. This also raises general concerns about the treatment of electricity consumption, an important part of household budgets in many countries, in poverty measurements. Where actual household data on consumption has been collected (directly from the utilities), consumption is very low, sufficient only to satisfy basic subsistence needs. In countries with very low consumption demand for electricity is relatively inelastic, suggesting that there may be large welfare losses associated with future tariff increases.

There may be substantial positive social benefits associated with access to electricity, suggesting an important continuing role for the public sector. This report is agnostic on the empirical evidence of the effectiveness of alternative instruments that mitigate the blow of tariff increases for the poor, such as income transfers and lifelines. It does, however, advocate helping clients calculate the social and fiscal implications of a full set of alternative mitigating strategies so that they can make fully informed choices.

---

35. A kink in the demand curve is a specific instance of non-linearity.

CHAPTER 5

LOCAL VERSUS GLOBAL ENVIRONMENTAL BENEFITS

Power sector reforms are expected to produce environmental benefits. Increased production efficiency, new investment, and environmentally friendly technology all contribute to lower fossil fuel consumption and lower emissions. With falling demand (the result of higher prices) this would lead to better ambient air quality and presumably to better health outcomes for the local population.

There may well be unanticipated environmental costs as well. The cold climate, rising price of electricity, and collapse of clean, safe heating alternatives (such as district heating) may push households, especially the poor, to substitute less expensive dirty energy (such as wood, coal, and kerosene) for electricity in heating. Burning dirty fuels lowers indoor and outdoor air quality, leading to worse health outcomes. Burning wood can also contribute to deforestation and the loss of valuable forest functions.

This chapter analyzes the environmental benefits and costs of reforms. It starts with a brief look at the expected environmental benefits of the reforms and whether they materialized. It then goes on to examine unintended costs, including fuel-switching behavior and potential health and deforestation outcomes.

Did the Reforms Achieve Environmental Benefits?

World Bank project documents outline specific environmental benefits from power sector reform (Box 5.1). These include a variety of emissions reductions and, in some cases, improvements in ambient air quality, especially around generation plants. Unfortunately, claims about improvements in ambient air quality are difficult to verify for most pollutants. Indicators for pollutants and monitoring programs were never established—or if they were, collection collapsed with the breakup of the Soviet Union. During the transition the collection of accurate air quality data decreased substantially in the countries of the Caucasus and Central Asia.

The environmental performance of the electricity sector, measured by fuel efficiency of electricity production, has improved slightly over the last decade, leading to reductions in carbon

Box 5.1: Reform Measures Expected to Result in Environmental Quality Improvements

- *Georgia:* promote efficient use and conservation of electricity, improve the efficiency of power production units, and implement a program to monitor ground-level concentration of total suspended particulates, sulfur dioxide, and nitrogen oxides for five years after commissioning.
- *Poland:* ensure proper consideration of externalities (environmental and other social costs) associated with energy production and use, and improve environmental performance by allowing the government to raise emission permit fees and penalties for exceeding the specified limits.
- *Moldova:* rehabilitate boilers and reduce consumption of imported fuel and gas by replacing deteriorated inefficient equipment, reducing emissions per kilowatt hour of produced electricity; the lower air emissions would reduce ground-level concentrations of these pollutants near the plants.
- *Armenia:* increase the use of natural gas and improve energy efficiency, reducing emissions.
- *Hungary:* increase production efficiency, reducing emissions and improving air quality.

Source: World Bank (1993a); World Bank (1993b); World Bank (1996); World Bank (1997)

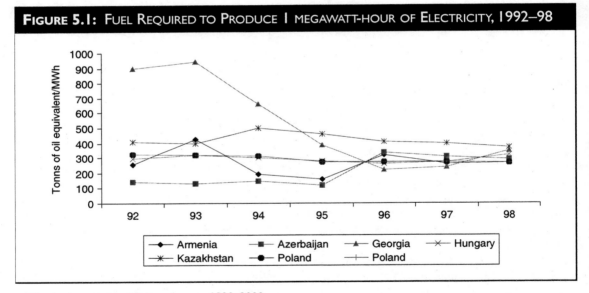

Figure 5.1: Fuel Required to Produce 1 Megawatt-Hour of Electricity, 1992–98

Source: International Energy Agency, 1990–2000.

dioxide emissions (Figure 5.1). Sophisticated climate change models, well beyond the scope of this study, are necessary to evaluate the benefits. Since these benefits are global rather than local, the incremental costs could be partially financed through global grant financing/emission trading and not be fully borne by the local population, which experiences only a small share of the benefits.

Increasing energy efficiency in electricity production does not automatically improve environmental health. Improvements resulting from increased efficiency can be insignificant when compared with emissions of sulfur dioxide, nitrogen oxides, and fine particulates (PM_{10}), the pollutants that cause the most health damage.[36] The health effects of particulate matter may be very

---

36. Different pollutants are associated with different health risks, commonly measured in terms of the disability adjusted life years (DALY), a measure used internationally to compare health effects of different causes. One DALY is equal to the loss of one healthy life year.

> **BOX 5.2: ESTIMATING THE POWER SECTOR'S CONTRIBUTION TO AIR POLLUTION AND HEALTH DAMAGE**
>
> A crude dispersion model was used to estimate the ambient air quality and consequent health damage from air pollution caused by the electricity sector in Almaty, Baku, Budapest, Katowice, Krakow, Tbilisi, Warsaw, and Yerevan. These were selected because in 2000 they were densely populated (750,000 or more inhabitants) and because exposure to primary combustion-related particles is usually higher in urban areas (UN 2002, 2003).
>
> The relationship between air pollution and health outcomes—such as mortality, chronic bronchitis, emergency room visits, and respiratory and cardiovascular hospital visits—is based on existing international studies. These studies examine the dose response relationships between exposure to particles (PM_{10}) and sulfur dioxide and health outcomes. (This is described in detail in Annex F.)
>
> Power sector emissions are simulated using a dispersion model that allocates a share of total ambient pollutant concentrations to the power sector. This allows a link to be established between ambient pollutant concentrations and power sector emissions, which can then be used to estimate health effects. (The base emission factors for major combustion processes used in the study are presented in Annex G.)

Source: Lvovsky (2002), authors' calculations.

low even if energy intensity is high if abatement equipment is used. If power plant stacks are high[37] or located in sparsely populated areas, as in many of the countries, they may not have much influence on ambient air quality, so increasing efficiency may not translate into local health benefits.

The raw data suggest that urban air pollution decreased slightly in the major cities during the reforms, yet it continues to be a major health hazard. Ambient standards for PM_{10} continue to be surpassed regularly in Yerevan, Tbilisi, and Katowice (Annex F).[38]

The key question is how much the power sector reforms contributed to this change. A crude dispersion model was used to estimate the magnitude of the impact of the sector on air quality and health in selected cities (Box 5.2 and Annex F). If the sector does not play an important role in determining local air quality, the reforms will produce small health benefits even if emission reductions are large. Conversely, if the sector plays an important role, the health benefits may be large.

Within the countries under study the power sector contributed less than 1 percent of health damage from all emissions. The share is small because the power stations account for a small amount of total ambient air quality. Between 1990 and 2000 the share of the electricity sector in the disability adjusted life years (DALYs) originating from low air quality ranged from 0.1 to 2.0 percent. The highest shares attributed to the sector are in Almaty and Warsaw. In the other cities the contribution is less than 0.5 percent.[39]

---

37. In the former Soviet Union a number of state norms and rules regulated the height and design of the chimneys of power plants. These rules and norms were in general close to western norms. The Ekibastuz Power Plant in Kazakhstan, which uses coal as fuel, has two stacks of 330 meters each. Other known stacks in Russia and Ukraine range from 250 meters to 1,370 meters.

38. Baku, Yerevan, and Tbilisi used to be included in the list of most polluted cities of the former Soviet Union, due to the industrialization and urbanization of the past 30 years. Lack of monitoring data precludes in-depth assessment of the state of the air quality, though air quality has been monitored in all the countries for many years. After decentralization, lack of funds and obsolete monitoring methods inhibited progress, and data collection has declined sharply.

39. Total DALYs from ambient air pollution range from around 4,000 on average in Krakow to around 50,000 on average in Katowice.

The analysis reveals five reasons for the low contribution of the electricity sector to health damages. First, the amount of electricity produced dropped substantially in Armenia, Georgia, and Kazakhstan. Second, the fuel mix used for thermal power plants shifted more toward natural gas in Armenia and Azerbaijan (Annex I). Third, high-capacity power plants are often located far from populated cities. Fourth, improvements in fuel quality[40] and abatement technologies for particulate matter were already in place before the reforms started in Hungary, Kazakhstan, and Poland, with average removal efficiencies of 97–99.9 percent. Fifth, power station stacks were built high to reduce deterioration of ambient air quality and were regulated by Soviet norms and regulations. (A detailed overview of the reasons for a low environmental health contribution of the power sector is provided in Annex H.)

Private transport is now a major source of urban air pollution in the large cities of Eastern Europe, the Caucasus,[41] and Central Asia. These emissions are increasing due to the aging vehicle fleet, the low quality and high sulfur content of the fuel, and the decline in the provision of public transport. In contrast, the share of emissions from power stations and other stationary sources is falling.

In sum, sector reforms are most closely associated with a reduction in carbon dioxide emissions, which produces global, not local, benefits. At the same time, sector reforms are unlikely to have any significant impact on ambient air quality related to sulfur dioxide, nitrous oxides, and fine particulate matter, the pollutants that cause the most local health damage.

There Were Unintended Environmental Costs

There may also be unintended environmental costs associated with the reforms—particularly from household fuel-switching. As residential tariffs are brought to cost-recovery levels, households may switch to less expensive but dirtier energy (such as wood, coal, or kerosene) contributing to indoor and outdoor air pollution. Health damage may be substantial, especially in densely populated urban areas where household chimneys are low and there is little opportunity for the pollution to disperse. Cumulative damage from household emissions may well exceed the benefits from reduced power plant emissions.

While there are no comprehensive data on household emissions, there is some evidence that some households, especially the poor, are more likely to use dirty fuels for heating. In Armenia 80 percent of households and 95 percent of poor households reported using alternative fuel sources to reduce reliance on electricity, primarily wood (60 percent) and/or gas (24 percent) (Lampietti and others 2001).

Burning dirty fuels can cause indoor air pollution. Worldwide, inhalation of smoke from combustion of solid fuels causes about 36 percent of lower respiratory infections, 22 percent of chronic obstructive pulmonary disease, and 1 percent of trachea, bronchus, and lung cancer (WHO 2002). It is also associated with tuberculosis, cataracts, and asthma, but the evidence here is weaker. Nearly 3 percent of disability adjusted life years worldwide are attributed to indoor smoke—2.5 percent for males and 2.8 percent for females (WHO 2002). The health threat depends on the local technology and ventilation. Unfortunately, little information is available on household technology, ventilation, or indoor air pollution levels.

Damage from Dirty Fuel Use May Be Large

Burning dirty fuels for heating is likely to be a significant source of urban/ambient air pollution. A recent note by UNEP (2002) indicated rising air pollution due to increased low temperature emissions, a large share of which is attributable to household heating. In Katowice, one of Central Europe's most severely polluted cities, the primary source of local air pollution

---

40. Sulfur and ash content of the coal; sulfur content of liquid fuel.
41. In Tbilisi, for instance, transport accounts for 80 percent of total air pollutants.

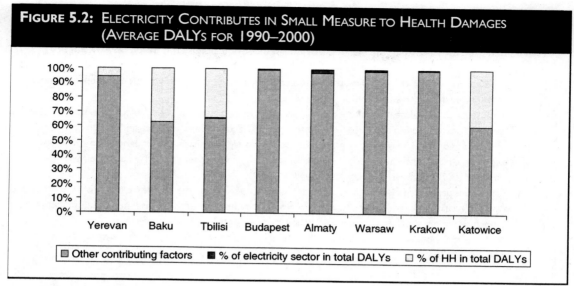

Source: Authors' calculations.

is household burning of coal for heating (Bucknall 1999; Bucknall and Gordon 2002; Gordon 2002).[42]

The dispersion model developed earlier, with assumptions about household fuel use, is used to estimate the share of ambient air pollution attributable to household wood and coal use.[43] The health effects are quantified only in those cities where, subject to data availability, it is reasonable to expect that the use of dirty fuels may rise following increases in electricity tariffs. The total share of DALYs attributable to households using "dirty" fuels ranges between 6 percent and 39 percent over the last decade (Figure 5.2), considerably higher than the contribution of the electricity sector to the DALYS originating from ambient air pollution (0.5 to 2.4 percent).[44]

Back-of-the-envelope estimates (Box 5.3) of the possible extent of health damage from indoor air pollution in three cities in the Caucasus puts the number of premature deaths at the same order of magnitude as that from outdoor air pollution (Table 5.1). More research is

---

42. In the Katowice area annual average levels of sulfur dioxide exceed the current EU standard nearly threefold, and annual average levels of fine particulates (PM_{10}) are well above the standard. Some parts of the metropolitan area exceed daily PM_{10} limit values for 200 days a year, causing significant respiratory illness and other problems for the population. The largest part of the average exposure to PM_{10} comes from household boilers, responsible for 80 percent of exposure to harmful particles in Katowice voivodship. Power and district heating plants contribute little to exposure because—owing to their high stacks—their emissions are dispersed over a wider area. The ambient air pollution impact in Katowice is attributable to coal and wood use for heating and cooking purposes, not solely as substitution for electricity.

In Katowice, 409 households burn coal for heating, and lower income households are more likely to heat with coal.

43. The following assumptions are made: 1) share of population using wood is as reported earlier, 2) the urban exposed population of Baku is estimated at 50 percent, 3) the average quantity of wood per household is 8 m^3 per household per year, 4) the density factor of wood is 0.5 ton/m^3, and 5) average household size is taken from UN World Prospects Population Database.

In the recently released results of the 2002 census of Georgia, the actual population appears to be smaller than originally listed in international databases of UN and World Bank. The contribution of the household sector to atmospheric air pollution comes down from 34 percent to 29 percent on average for the last 10 years. The relative contribution of the electricity sector to health effects will accordingly be lower, since it also depends on the number of people affected.

44. These are total and not incremental figures.

> **Box 5.3: Methodology for Calculating Damages from Indoor Air Pollution**
>
> The method used for calculating the health burden from indoor air pollution is based on Kirk Smith's approach[45]:
> - Estimate the size of the population exposed to indoor air pollution from wood and coal use and the absence of sufficient ventilation or good stoves.[46]
> - Assess the potential health risk factors, based on existing studies in South Asia, Latin America, and Sub-Saharan Africa, specifically focused on women and children under five. This underestimates the effect on males and the elderly since these studies focus mainly on health consequences from cooking. The exposure of men and the elderly is higher in the case of dirty fuel use for heating than when women usewood and coal mostly for cooking only.
> - Assess the pattern of risk diseases, using statistics to determine the current pattern of diseases associated with indoor air pollution: acute upper respiratory infection, bronchitis, pneumonia, influenza, bronchitis, emphysema and asthma and other diseases of the respiratory system.
> - Calculate the potential disease burden in DALYs related to indoor air pollution if there are no chimneys or ventilation.

Source: Smith (2000).

Table 5.1: Exposure to Indoor Air Pollution is High

	Armenia (Yerevan)	Georgia (Tbilisi)	Azerbaijan (Baku)
Number of premature deaths			
Children under five	52 (19)	62 (20)	36 (17)
Women	164 (60)	147 (47)	114 (54)
Total	216 (79)	210 (66)	150 (71)
DALYs			
Children under five	1,820 (664)	2,186 (690)	1,260 (597)
Women	3,287 (1,199)	2,928 (931)	2,275 (1,078)
Total	5,107 (1,863)	5,134 (1,621)	3,535 (1,675)

Source: Authors' calculations based on WHO statistics, mortality database, and household surveys.

necessary to identify the relationships between fuel use (including technology and chimney availability) and indoor air pollution and health outcomes.

The number of premature deaths is higher among women than children under five, counterintuitive given the evidence in other continents. Yet, in ECA the number of women compared with the number of children is very different from that in other continents: in Azerbaijan the ratio of women to children is 6, in Georgia 10, and in Armenia 8.5.

The total estimated potential loss of life due to indoor air pollution amounts to 7 percent of all deaths related to respiratory diseases and 1 percent of the total deaths in Armenia, 10 percent

---

45. Based on an estimated exposed population of 50 percent for Armenia, 57 percent for Georgia, 30 percent for Azerbaijan, and an odds ratio of 2 for all countries.

46. The estimation of indoor air pollution effects due to absence of good ventilation and stoves is indicative since not enough information is available on wood use cooking and heating methods. Specific research is needed to establish reliable odds ratios.

of all deaths related to respiratory diseases and 1 percent of the total deaths in Georgia, and 2 percent of all deaths related to respiratory diseases and 0.3 percent of total deaths in Azerbaijan.

Fuel wood use may also contribute to deforestation and the loss of important forest resources. However, it is unclear whether this is a problem because data on deforestation, particularly deforestation attributable to fuel wood collection, are notoriously difficult to obtain. Several studies (such as UNEP 2002) and observations of forestry specialists visiting the Caucasus show a significant decrease of local forest cover and a deterioration in forest quality. But these trends are often not reflected in national or international statistics. The United Nations Economic Commission for Europe and Food and Agriculture Organization report (2000) on forest resources of Europe does not indicate a decline in forest resources.[47] Indeed, the forested area appears to be increasing from 0.2 percent a year in Poland to 2.2 percent a year in Kazakhstan in the last 10–15 years.[48] Yet, these data are unlikely to be reliable. Few ministries have the resources to monitor forest cover consistently and rigorously.

Changes in forest canopy cover, which can be monitored using conventional remote sensing approaches, often do not reflect changes in forest health, yield, species mix, or density, which can be captured only by more rigorous ground based inventories and assessments. The low-intensity harvesting of fuelwood from trees growing in agricultural land, around houses, and along roads is seldom shown to have a significant impact on overall forest canopy cover—and is difficult to measure with remote sensing. Particularly when trees are coppiced or pollarded to provide these supplies, the overall impact of rural firewood harvesting can be negligible.

The situation is much different in meeting urban demands for firewood. Heavy urban household use of fuelwood creates market conditions that favor the clearance of sometimes large forested areas. When urban household energy use is constrained as a result of utility reform, the negative environmental impact on forested areas can be significant because of a shift from electricity to firewood. It may also happen that there is no visible change in total forest cover, even though the quality and density is decreasing (UNEP 2002).[49]

From the household surveys it is clear that the majority of rural households and a substantial number of urban households used wood for their energy needs. With average expected consumption of 5–10 cubic meters of fuelwood per year, this can lead to substantial (local) deforestation. More research is necessary on the amounts of fuelwood that households burn, the sustainability of this practice, and the incremental use of fuelwood due to electricity reforms.

Power generation can also damage forest cover. Air pollutants originating from power plants (specifically sulfur dioxide) damage forests through acidification (and the well known acid rain). The precise impact of power sector emissions on forests and the impact of the power sector privatization on these forests were not analyzed in this study, since the focus was on poverty-environment linkages.

Conclusions

Power sector reforms are expected to produce environmental benefits. Increased production efficiency, new investment, and environmentally friendly technology all contribute to lower fossil

---

47. A forest is defined as a forest when composed primarily of indigenous (native) tree species. Natural forests include closed forests and open forests (where at least 10 percent tree cover). Total forests consist of all forest area (plantations and natural forests) for temperate developed countries.
48. Reference periods for the different countries are Armenia 1983–96, Azerbaijan 1983–88, Kazakhstan 1988–93, Poland 1987–91 and 1992–96, Hungary 1990–96, and Moldova 1990–95. There is no information available for Georgia.
49. The Caucasus Environmental Outlook reports that selective cutting occurred, when the highest quality trees were cut. During the last 10 years, cutting was extensive on the Saguramo-Yalon range (East Georgia), and on the outskirts of Tblisi and Yerevan. In state-owned forests, there were no significant changes in the total cover, but all valuable specimen of beech and some other species have been cut, drastically reducing forest quality. It is estimated by the Caucasus Environmental Outlook that 26 percent of beech forest has been converted to coppice forests and only about 10 percent of the beech forests left have a high density in Armenia, and Georgia.

fuel consumption and lower emissions. This leads to better ambient air quality and presumably to better health outcomes for the local population.

Unfortunately, claims about improvements in ambient air quality are difficult to verify because monitoring programs were never established. Better monitoring of ambient environmental quality improvements is necessary in the future. For the long-term sustainability of reforms, it is critical that environmental impacts be systematically followed and recorded. It is appropriate to define mitigation measures for negative environmental impacts, such as those on health. Impact analysis should be on a national scale rather than from a sector perspective.

Reforms did slightly improve energy efficiency in power plants. This matters for carbon dioxide emissions and global climate impact. In most cases this has little direct impact on human health since the electricity sector's share of the total health damage from air pollution is negligible.

It is also possible that reforms have damaged health because households switched to dirty fuels. Household fuel burning has a high impact on ambient urban air quality because households often live in densely populated areas in houses with low chimneys and no pollution abatement equipment. As noted in the previous section, a key solution is improving access to and efficiency in the use of clean alternatives. Survey data indicate that fewer households would use wood and coal if they had access to gas. Of course, in many countries the gas sector is also in need of reform before it can operate on a sustainable basis.

As in the previous section, household surveys do not reveal enough information about energy (and other utility) reforms. More data are needed to evaluate the impact of reforms on fuel switching, energy use, substitution effects, and health and social impacts. This calls for including questions about utilities in such surveys and developing models to help predict behavior under a variety of scenarios (suggested questions are presented in Annex J).

ANNEX A

OVERVIEW OF THE REFORM PROCESS IN EIGHT ECA COUNTRIES

	Regulatory Development	Corporatization and Unbundling of Monolithic Company	Privatization of Distribution	Privatization of Generation
Armenia	1997: Energy Law established an independent Energy Commission	In 1997 Armenergo was unbundled	Privatization negotiations with strategic investors are on-going.	No privatization has yet taken place.
Azerbaijan		Following the approval of the "Azerbaijan Republic Law on electric power engineering" (1998), the Power Grid was divided into three parts: the State electric energy enterprise, independent power producers and power supply enterprises.		In 2000–2001, management of established Power generation enterprises was contracted to for the 25 years period to two private companies—Turkish "Barmek Holding" and to JSC "Baku High-voltage Equipment". First steps toward privatization of generation have been made. In particular, a Presidential Decree of December 21 2001 "On privatization of small Hydroelectric Power Plants" was approved and privatization of nine small HPPs has started.
Georgia	1997: Electricity Law established an independent regulator Georgia National Energy Regulatory Commission (GNERC). Wholesale Electricity Market was established in 1999.	In 1999–2000 Sakenergo was unbundled.	1998: distribution in Tbilisi (its share is 30–50 percent of total national consumption) was sold to AES Power of the USA. In 2003 AES sold its stake in Georgia energy business to RAO UES of Russia.	In 2000 2 units (of a total of 8, of which 6 are not operational) at the Thermal generation plant Tbilsresi were sold to AES. AES also managed the Khrami hydro generation station.

(Continued)	Regulatory Development	Corporatization and Unbundling of Monolithic Company	Privatization of Distribution	Privatization of Generation
Hungary	1993: Policy guidelines 1994: Electricity Act; Established in 1994, Hungarian Energy Office (HEO) is the regulatory and supervisory body for gas and electricity companies, heat production of electric power stations and CHP companies over 50 MW, and protecting consumer interests.	1993–94: MVM Trust unbundled into 8 generation companies, one transmission utility, and 6 distribution companies (EDC)	1995: controlling shares in each of 6 EDCs sold to strategic investors (mainly German and French), raising about US$ 1.1 bln in revenues. 1997–98: remaining shares in EDCs are sold through stock market offering	1995: controlling shares in 2 generation companies sold to strategic investors; 1996–97: 4 more generation companies; and 1 more were privatized later. All power stations have been privatized except the nuclear and an old coal-fired station.[50]
Kazakhstan	1996: a series of government resolutions to unbundle; 1998–99: Law on Natural Monopolies, Law on Electricity and creation of the regulatory Anti-Monopoly Agency (AMA)	1996: unbundling of Kazakhenergo	Since 1996, 3 out of 18 distribution companies have been privatized (electricity and heat distribution networks in the Almaty region to Tractabel of Belgium in 1996; electricity networks in Karaganda region to National Power of UK in 2000; the networks in the Altai region to AES Power of USA in 1999.	Since 1996, around 80 and 90 percent of generation assets have been privatized. The government is believed to have privatized the remaining generation assets to RAO UES of Russia in 1999–2002.
Moldova	1998: Electricity law was approved in 1998 and established an independent regulatory agency ANRE	1997: Moldenergo was unbundled into 3 generation companies, 5 distribution, and 6 other construction and heat companies and a state enterprise responsible for transmission and dispatch	1999: sale of 3 out of 5 distribution companies (covering more than 2/3 of the market) to Union Fenosa of Spain	No privatization has yet taken place.

(Continued)

	Regulatory Development	**Corporatization and Unbundling of Monolithic Company**	**Privatization of Distribution**	**Privatization of Generation**
(Continued)				
Poland	1997: Energy Law laid out reforms and created an independent Energy Regulatory Agency (ERA)	1993: Commercialization and unbundling of PSE (Polish Power Grid Company)	In 2003, five distribution companies in Western and Northern Poland were consolidated (P-5). Future plans include creating three more power distribution enterprises. Privatization is underway in eight power distribution enterprises in Northern and Central Poland. Together, they command 16% market share[51].	Consolidation of the generation sector continues with merger of PKE SA and BOT. Plans are underway to merge five other companies that would constitute 26% of national installed capacity. No privatization has yet taken place.

50. http://www.uccee.org/SectorReform/ImpactsHungary.pdf
51. http://www.warsawvoice.pl/view/4892

Sources: (1) Private Sector Participation in the Power Sector in ECA Countries: Lessons from the Last Decade. World Bank. 2002. Draft; (2) Privatization of the Power and Natural Gas Industries in Hungary and Kazakhstan. World Bank. 1999; (3) news sources.

ANNEX B

PROCEEDS FROM PRIVATIZATION OF ELECTRIC UTILITY COMPANIES

Armenia	Starting Nov. 1, 2002, Midland Resources British Company, registered off-shore, is the 100 percent owner of Armenia's Electric Power Nets. It paid $12.015 million for 100 percent stocks and has assumed liability to redeem debts in the amount of $27.985 million (of which $22 ml are debts to Armenian commercial banks).
Georgia	AES of the US paid 25.5 million dollar equivalent in GEL for a 75 percent stake in Telasi in 1999. In addition, AES agreed to pay $10 millon of existing Telasi debt. It also committed to invest 83.8 m dollars within the next 10 years and provide consumers with 24-hour electricity.
Hungary	Privatization of electric companies through investors raised significant revenues: some $1.5 billion, of which some $150 ml went to MVM Rt (the former monolithic state owned power company). Most companies were profitable at the time of purchase.
Kazakhstan	The sale did not raise any significant revenues for the government. The exact amounts investors paid for electric companies remained confidential; it is known from interviews that the sale price did not exceed $5 million for any one asset (except land); the investor paid the local government for land on which the asset stands in a separate transaction. Because of the weakness of regulation and the poor condition of the assets, the sale price was low. (1) Tractebel is supposed to have paid about $4 ml for the distribution concession, some generation assets and district heating assets. In April 1999 the government froze utility prices, which made the continued operation of Tractebel impossible, and it withdrew. (2) AES Silkroad reportedly paid $11 ml for the land on which Ekibastuz 1 stands to the regional government and a further $1.5 ml for the generation asset to the State Property Committee.
Moldova	A total of 17,259,826 shares in 3 regional electricity distribution companies were sold for $21.095 million to Union Fenosa in February 2000. Adding the price of public patrimony, the state received $25.3 million in total. Union Fenosa pledged to invest $67 million over a five-year period. According to PWC, it has so far met its investment obligations.
Poland	German energy company RWE AG paid 1.5 billion PLN for an 85 percent stake in Warsaw-area EDC, STOEN SA. Agreement was reached in October 2002 and was cleared by the authorities in December.

Sources: Privatization of the Power and Natural Gas Industries in Hungary and Kazakhstan. The World Bank Technical Paper No. 451. December 1999; various news sources.

Annex C

Tariff Losses, Commercial and Collection Losses, as Share of Total Losses

	1995	1996	1997	1998	1999	2000
Tariff losses, percent of total						
Armenia	26	21	11	14	16	18
Azerbaijan	63	70	55	57	64	64
Georgia	12	19	15	14	9	10
Hungary	100	100	100	100	100	100
Kazakhstan	41	44	Na	na	na	20
Poland	100	100	100	100	100	100
Commercial losses, percent of total						
Armenia	46	42	36	42	54	54
Azerbaijan	11	6	11	1	1	2
Georgia	62	66	66	66	68	68
Hungary	na	na	na	na	na	na
Kazakhstan	19	30	na	na	na	15
Poland	na	na	na	na	na	na
Collection Losses, percent of total						
Armenia	28	37	53	44	30	28
Azerbaijan	26	24	35	42	35	34
Georgia	26	16	19	20	23	22
Hungary	na	na	na	na	na	na
Kazakhstan	40	26	na	na	na	65
Poland	na	na	na	na	na	na

Annex D

Fiscal Balance and Electricity Sector Financial Deficit (Million USD and Share of GDP, Respectively)

	1995	1996	1997	1998	1999	2000
Overall Deficit/Surplus ($ million)						
Armenia	116 (9%)	136 (8.5%)	95 (5.8%)	79 (4.2%)	98 (5.3%)	89 (4.6)
Azerbaijan	118 (4.9%)	89 (2.8%)	63 (1.6%)	173 (3.9%)	214 (4.7%)	
Georgia		215 (7.1%)	220 (6.1%)	177 (4.9%)	139 (5%)	77 (2.6)
Hungary	1418 (3.2%)	368 (−0.8%)	824 (1.8%)	2949 (6.3%)	1655 (3.4%)	1557 (3.4%)
Kazakhstan	649 (3.9%)	1130 (5.4%)	1579 (7.1%)	1691 (7.6%)	843 (5%)	138 (0.8%)
Poland						
Electricity Sector Financial/Deficit						
Armenia	138 (10.7%)	128 (8%)	116 (7.1%)	80 (4.2%)	58 (3.1%)	56 (2.9%)
Azerbaijan	435 (18%)	383 (12%)	353 (8.9%)	418 (9.4%)	498 (10.9%)	556 (10.6%)
Georgia	284	258 (8.5%)	265 (7.4%)	288 (7.9%)	283 (10.1%)	266 (8.8%)
Hungary	491 (1.1%)	365 (0.8%)	160 (0.3%)	86 (0.2%)	67 (0.1%)	−46 (−0.1%)
Kazakhstan	1061 (6.4%)	898 (4.3%)	687 (3.1%)	−139 (−0.6%)	416 (2.5%)	665 (3.6%)
Poland	97 (0.1%)	56 (0.0%)	407 (0.2%)	302 (0.2%)	322 (0.2%)	333 (0.2%)

Source: Financial deficit data provided by PREM; electricity sector financial deficit calculated in this study.

Annex E

Efficiency Indicators

	Units	1990	1991	1992	1993	1994	1995	1996	1997	1998	1999	2000	2001	2002
ARMENIA														
Available vs installed capacity	%	NA	67	67	67	67	81	81	81	81	81	81	81	NA
Net supply	GWh	NA	10,367	8,821	6,080	5,416	5,183	5,753	5,514	5,362	5,075	5,102	4,981	4,828
Sales	GWh	NA	8,956	7,649	5,104	3,218	3,054	3,802	3,859	3,794	3,570	3,590	3,479	3,400
Sales—Households	GWh	NA	1,896	3,099	2,409	1,893	1,725	1,608	1,720	1,458	1,279	1,234	1,186	1,221
Metered consumption	GWh	NA	8,956	7,649	5,104	3,218	3,054	3,802	3,859	3,794	3,570	3,590	3,479	3,400
Technical losses	% of net supply	NA	14	13	14	14	16	14	14	16	17	17	18	16
Commercial losses	% of net supply	NA	0	7	16	27	25	20	16	13	13	12	13	13
Collection rate	% of metered consumption	NA	100	50	50	39	54	60	61	77	88	89	81	90
Reported cost of generation	US cents/kWh	NA	NA	NA	NA	1.5	1.7	2	2.6	2.3	2.7	2.4	2.3	1.5
Average tariff	US cents/kWh	NA	2.0	0.8	0.6	2.7	2.7	3.1	4.1	4.0	4.1	4.0	3.8	3.7
Average tariff—Households	US cents/kWh	NA	2.0	0.1	0.2	2.7	2.0	2.4	3.4	3.8	4.7	4.6	4.5	4.3
Number of employees	People	NA	8,692	8,728	9,254	10,398	12,353	16,520	17,563	19,289	18,062	17,444	16,311	16000
Salary per employee (power sector total)	US $	NA	110	1	20	62	47	59	61	71	93	95	99	98
Customers per employee		NA	100	99	81	72	61	45	43	39	42	43	46	49
Sales per employee	MWh/year	NA	1,030	876	552	309	247	230	220	197	198	206	213	213
AZERBAIJAN														
Available vs installed capacity	%	90	83	90	87	84	73	70	71	72	72	73	72	58
Net supply	GWh	20,281	20,485	17,960	17,999	13,756	12,945	13,246	12,788	14,263	15,003	15,112	16,162	16,095
Sales	GWh	NA	17,507	15,290	14,566	13,644	12,974	13,276	12,968	14,263	15,002	15,608	16,162	18,031
Sales—Households	GWh	NA	2,875	2,984	3,776	4,663	5,092	6,119	5,819	7,481	9,229	9,965	10,232	10,540
Metered consumption	GWh	17,190	17,823	15,487	14,686	10,806	9,560	10,069	9,147	11,352	12,217	12,147	13,874	13,118
Technical losses	% of net supply	15	12	13	14	15	18	20	22	20	18	18	12	13
Commercial losses	% of net supply	1	1	1	5	7	8	4	6	0	0	1	2	6
Collection rate	% of metered consumption	89	93	78	62	35	39	57	50	43	34	16	26	34
Reported cost of generation	US cents/kWh	3.0	1.1	2.8	3.0	0.6	1.9	2.2	2.5	2.3	1.9	1.8	1.8	1.5
Average tariff	US cents/kWh	3.4	2.0	0.4	1.4	0.7	2.0	2.1	2.7	2.7	2.2	1.9	1.5	1.9
Average tariff—Households	US cents/kWh	6.7	1.1	0.2	0.3	0.2	0.4	1.2	2.4	2.5	2.3	2.1	2.1	2.0
Number of employees	People	NA	17,951	20,709	21,343	21,078	21,026	20,805	21,554	21,823	21,387	20,645	20,115	18,826
Salary per employee (power sector total)	US $	NA	75	30	42	38	38	46	67	76	66	75	107	105
Customers per employee		NA	75	65	64	67	70	73	72	71	73	76	78	81
Sales per employee	MWh/year	NA	975	738	682	647	617	638	602	654	701	756	803	958

	Units	1990	1991	1992	1993	1994	1995	1996	1997	1998	1999	2000	2001	2002
GEORGIA															
Available vs installed capacity	%	NA	42	42	42	42	42	42	42	42	42	NA	NA	NA
Net supply	GWh	NA	NA	11,520	10,150	7,044	7,082	7,232	7,172	8,062	8,098	7,447	6,905	7,215
Sales	GWh	NA	NA	4,638	3,948	2,761	2,703	2,466	2,539	2,783	2,800	2,635	2,510	2,602
Sales—Households	GWh	NA	NA	1,043	1,386	1,217	1,140	1,028	1,106	1,896	1,926	1,819	1,676	1,738
Metered consumption	GWh	NA	NA	4,638	3,948	2,761	2,703	2,466	2,539	2,783	2,800	2,635	2,510	2,602
Technical losses	% of net supply	NA	NA	13	13	13	13	15	15	15	15	15	15	15
Commercial losses	% of net supply	NA	NA	50	50	50	50	50	50	50	50	50	50	50
Collection rate	% of metered consumption	22	92	63	28	20	22	42	38	39	40	42	45	47
Reported cost of generation (HYDRO)	US cents/kWh	NA	NA	NA	NA	NA	NA	NA	NA	0.4	0.5	0.7	0.7	0.7
Reported cost of generation (THERMAL)	US cents/kWh	NA	NA	NA	NA	NA	NA	NA	NA	3.7	4.3	3.1	2.3	2.7
Reported cost of generation (WEIGHTED AVERAGE)	US cents/kWh	NA	NA	NA	NA	NA	NA	NA	NA	1.1	1.3	1.2	1	0.9
Average tariff	US cents/kWh	NA	2.0	2.0	0.0	0.0	3.5	2.9	3.2	3.4	3.9	3.8	3.9	4.3
Average tariff—Households (Tbilisi)	US cents/kWh	NA	4.0	4.0	4.0	4.0	3.5	2.0	2.6/3.6	3.40	4.70	4.6/5.0	4.8/6.0	6.6
Number of employees (power sector total)	People	NA	21,000	NA	NA	NA	NA	NA	NA	17,844	17,750	17,900	18,000	18,000
Salary per employee	US $	NA	NA	NA	NA	NA	NA	NA	208	115	132	151	173	177
Customers per employee		NA	64	NA	NA	NA	NA	NA	NA	64	64	63	63	63
Sales per employee	MWh/year	NA	NA	NA	NA	NA	NA	NA	NA	156	158	147	139	145

NOTE: Commercial losses are based on expert opinion.

	Units	1990	1991	1992	1993	1994	1995	1996	1997	1998	1999	2000	2001	2002
HUNGARY															
Available vs installed capacity	%	95	93	92	89	91	92	94	92	93	94	94	94	NA
Net supply	GWh	37,036	34,816	32,586	32,828	32,993	33,668	34,554	34,581	34,998	35,285	35,884	35,872	NA
Sales	GWh	33,010	30,945	29,745	28,470	28,470	28,919	29,877	29,847	30,082	30,455	31,151	32,196	NA
Sales—Households	GWh	9,189	9,768	10,514	9,721	9,842	9,787	10,053	9,780	9,769	9,838	9,792	10,130	NA
Metered consumption	GWh	33,010	30,945	29,745	28,470	28,740	28,919	29,847	29,847	30,082	30,455	31,151	32,196	NA
Technical losses	% of net supply	11	11	9	13	13	14	14	14	14	14	13	13	NA
Commercial losses	% of net supply	0	0	0	0	0	0	0	0	0	0	0	0	NA
Collection rate	% of metered consumption	100	100	100	100	100	100	100	100	100	100	100	100	NA
Reported cost of generation	US cents/kWh	NA	NA	NA	NA	NA	NA	NA	NA	NA	NA	NA	NA	NA
Average tariff	US cents/kWh	3.6	5.7	4.8	4.4	3.5	3.3	3.7	4.3	4.6	4.6	4.9	5.0	NA
Average tariff—Households	US cents/kWh	2.0	3.1	3.6	3.7	3.0	4.3	4.6	5.4	5.7	5.9	6.6	NA	NA
Number of employees (power sector total)	people	NA	38,067	37,063	40,521	44,746	43,693	41,990	40,203	39,636	34,988	31,490	27,142	NA
Salary per employee	US $	NA	NA	NA	NA	NA	NA	NA	NA	NA	NA	NA	NA	NA
Customers per employee		NA	NA	NA	NA	NA	NA	120	126	128	146	163	189	NA
Sales per employee	MWh/year	NA	813	803	703	642	662	711	742	759	870	989	1,186	NA

	Units	1990	1991	1992	1993	1994	1995	1996	1997	1998	1999	2000	2001	2002
KAZAKHSTAN														
Available vs installed capacity	%	84	87	87	84	85	80	79	75	73	74	74	76	76
Net supply	GWh	97,450	94,300	89,450	81,940	73,000	67,740	59,990	50,620	47,160	44,640	48,740	50,770	52,141
Sales	GWh	89,250	86,170	80,600	72,750	63,430	57,520	48,900	40,700	37,350	35,440	42,180	43,960	NA
Sales—Households	GWh	7,330	8,060	8,330	9,110	9,050	8,100	7,150	6,120	5,520	5,650	5,490	5,110	NA
Metered consumption	GWh	89,250	86,170	80,600	72,750	63,430	57,520	48,900	40,700	37,350	35,440	42,180	43,960	NA
Technical losses	% of net supply	8	8	8	8	8	8	7	7	7	7	7	8	NA
Commercial losses	% of net supply	0	0	2	3	5	7	11	12	14	14	7	5	NA
Collection rate	% of metered consumption	100	100	87	89	99	77	85	63	82	74	62	62	NA
Reported cost of generation	US cents/kWh	0.7	NA	NA	1.4	2.2	2.6	3.2	NA	NA	2.5	2.2	1.7	NA
Average tariff	US cents/kWh	1.5	2.0	2.0	1.3	2.1	3.2	3.2	4.6	4.8	3.2	2.7	2.6	NA
Average tariff—Households	US cents/kWh	4.0	2.0	1.0	0.7	2.4	3.0	3.0	2/3.8	3/4.2	0.50	2/3.6	2/3.4	2/3.2
Number of employees (power sector total)	people	NA	74,100	83,600	85,300	91,400	94,300	91,300	96,100	144,500	137,200	135,400	139,300	NA
Salary per employee	US $	NA	NA	NA	NA	NA	NA	NA	NA	NA	NA	NA	NA	NA
Customers per employee		NA	NA	NA	NA	NA	NA	NA	NA	NA	NA	NA	NA	NA
Sales per employee	MWh/year	NA	1,163	964	853	694	610	536	424	258	258	312	316	NA
POLAND														
Available vs installed capacity	%	NA	90	91	91	91	91	91	90	90	91	90	90	NA
Net supply	GWh	NA	NA	NA	NA	NA	NA	NA	NA	NA	NA	NA	NA	NA
Sales	GWh	NA	96,384	91,586	92,344	93,378	96,119	99,275	101,095	101,105	99,376	100,506	97,980	NA
Sales—Households	GWh	NA	19,326	18,429	18,206	18,207	18,075	19,224	19,771	20,314	20,800	21,037	21,376	NA
Metered consumption	GWh	NA	NA	NA	72,582	73,776	76,636	78,612	79,961	79,455	77,341	78,242	75,354	NA
Technical losses	% of net supply	NA	12	13	15	14	12	14	13	12	11	10	11	NA
Commercial losses	% of net supply	NA	0	0	0	0	0	0	0	0	0	0	0	NA
Collection rate	% of metered consumption	NA	100	100	100	100	100	100	100	100	100	100	100	NA
Reported cost of generation	US cents/kWh	NA	1.9	1.9	1.9	2.3	2.6	2.7	2.6	2.9	3	2.9	3.1	NA
Average tariff	US cents/kWh	NA	NA	NA	4.0	4.2	4.7	4.7	4.3	4.4	4.4	4.4	5.3	NA
Average tariff—Households	US cents/kWh	NA	NA	2.9	4.1	4.5	4.6	5.8	5.8	5.3	5.5	5.3	5.3	NA
Number of employees (power sector total)	people	NA	114,899	116,887	117,012	118,093	117,147	114,849	110,027	110,949	107,532	101,931	100,075	NA
Salary per employee	US $	NA	204	293	342	409	503	549	534	563	668	685	792	NA
Customers per employee		NA	124	124	125	125	126	129	136	136	141	150	154	NA
Sales per employee	MWh/year	NA	839	784	789	791	820	864	919	911	924	986	979	NA

Annex F

More on the Methodology for Estimating Health Effects

The major factor that determine the contribution of power sector the air pollutions that cause health damages are the fuel mix that is used by the power plants and the abatement technologies that are installed to reduce emissions. The amount of fossil fuels (coal, fuel oil, lignite) that is used by the power sector as well as the fuel quality (sulfur and ash content of the coal, sulfur content of liquid fuel) determine the base emissions of the power plants. These base emissions are calculated on the basis of standard emission factors and are presented in Annex B (WHO 1989, Lvovsky 2000).

The emissions coming from the power sector are then modeled in a crude dispersion model to allocate the share of the power sector in the total ambient air concentration of pollutants. This is done to establish a link between the change in ambient air concentrations of pollutants and the change of emissions from the power sector. The change in ambient air concentrations is used to estimate the health effects coming from the power sector.

Estimations for the quantitative effect of air pollution on mortality, cases of chronic bronchitis and lesser health impacts such as respiratory hospital admissions, cardiovascular hospital admissions, emergency room visits etc. are based on international studies examining the dose response relationships between the exposure to fine particles and to a lesser extent to exposure of SO_2. The dose-response functions used are presented in the following table.[52] These present a change in crude mortality rates and DALYs attributable to a change in 10 ug/m^3 of annual mean concentrations of PM_{10} and SO_2.

---

52. See Lvovsky (2000) and B. Ostro, "Estimating the health effects of air pollution," World Bank, 1994.

Health effects	Unit	PM_{10}	SO_2	DALYs per 10,000 Cases
Mortality	(percent change in the crude mortality rate)	0.084		100,000
Chronic bronchitis	per 100,000 adults	3.06		12,037
Respiratory hospital admissions	per 100,000 population	1.2		264
Emergency room visits	per 100,000 population	023.54		3
Restricted activity days	per 100,000 adults	05,750		3
Lower respiratory illness in children	per 1000 children	169		3
Respiratory symptoms	per 100,000 adults	18,300		3
Cough days	per 100,000 children		1.81	3
Chest discomfort days	per 100,000 adults		1,000	3

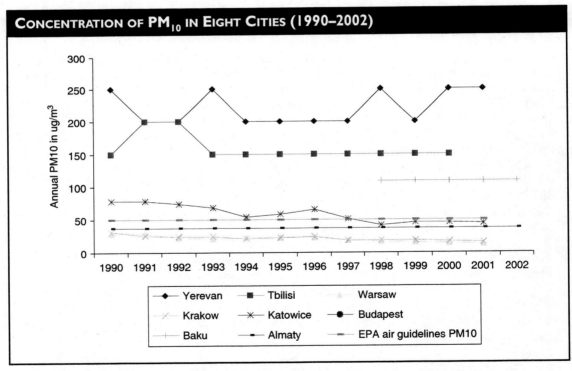

Source: Local Consultants Reports, UNEP (2002), Ongoing estimations of Pandey (World Bank, 2003)

Because PM_{10} is in most countries not measured directly, the figures are estimated from TSP measurements based on a conversion factor of 0.5–0.6 PM_{10} per TSP unit. The analysis has been focused on the major urban agglomerations in the countries, with at least 750,000 inhabitants or more in 2000 and where urban air pollution is likely to occur.

ANNEX G

BASE EMISSION FACTORS

The base emission factors for major combustion processes used in the study are presented in the following table.

Emission factor (kg per ton fuel)	TSP	PM_{10}	SO_2	NO_2
Coal	5*A	0.5–0.8 * $TSP_{em.f.}$	19.5 * S	10
Fuel Oil	2* (0.4 + 1.32)*S	0.6–0.9* $TSP_{em.f}$	20 * S	8.5
Lignite	3.1 *A	0.5–0.8 * $TSP_{em.f.}$	15 * S	6

A = Ash content of coal, weight percent
S = Sulfur content of fuel, weight percent
Source: WHO 1989 and Lvovsky 2000.

When the equipment is not new and not well maintained and monitored, an emission increase by factor 2–3 is applied.

Emission ratios between PM10 and TSP range between 0.5–0.8 for coal and 0.6–0.9 for fuel oil dependent on the removal efficiency for TSP control at power plants. The higher the removal efficiency, the higher the ratio for the amount of PM_{10} per TSP emission

ANNEX H

MORE ON FACTORS LEADING TO LOW CONTRIBUTION OF POWER SECTOR TOWARD HEALTH DAMAGES

Amount of electricity produced: Armenia experienced a reduction in thermal electricity production of 72 percent over the past 10 years, Georgia 76 percent, Kazakhstan 42 percent[53] and hence emissions also fell substantially. The development of the amount of electricity produced is graphically presented in Annex D.

Azerbaijan and Kazakhstan have more or less retained its industrial capacities and power sector. Poland and Hungary increased their power generation. In the Caucasus region, transport and industry were the major sources for air pollution during the 1970s and 1980s, but total emissions fell early 90s due to economic decline.

Armenia and Georgia suffered the most, where the share of stationary sources dropped to almost 5 percent of total emissions due to the power crisis. The emissions from power stations declined due to reduction in production rather then improved energy efficiency.

Fuel mix, fuel quality, location of power plants, abatement technologies: The fuel mix of the thermal power plants in **Azerbaijan** consists of Mazut (Heavy Fuel Oil) and Natural Gas. Overall the use of Mazut has been declining by 65 percent between 1990 and 2002 while the use of natural gas increased by 155 percent in the same period. While the Sulfur content has been declining from 0.56 percent in 1992 to 0.33 percent in 2002, the most important factor in the contribution of the power sector to health impacts consists of the location of the different power plants. Around 55 percent to 65 percent of the total Mazut used in the country was used in the power plant located near Mingechevir, located on the northwest part of Azerbaijan containing only around 100,000 inhabitants. Ali-Bayramli, responsible for 23–36 percent of total Mazut use, hosts only around 65000 inhabitants during the 90s.

The fuel mix of the thermal power plants in **Poland** consists of coal and lignite. Fuel oil is only used as a back up or starting fuel and therefore only consumed in very small quantities (less then 1 percent of total fuel consumption). The average ash content of the coal declined from

---

53. IEA. Various years. Energy Statistics of OECD-countries and Energy Statistics of non-OECD countries

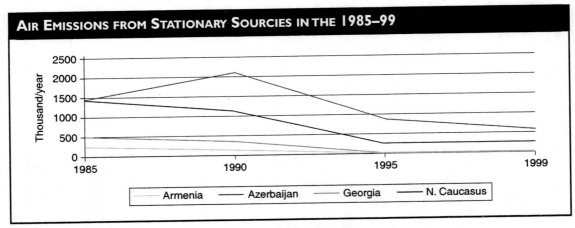

Source: Complex Scheme of Environment Protection Azerbaijan, 1987.
State Cimmitee on Ecology and Council of Natural Resources Utilization, Azerbaijan, 1993.
MoE of Georgia, 1996. State Stastistical Services of Armenia, Azerbaizan and Georgia, 1986–2000.

around 22 percent in 1992 to 20.5 percent in 2001 and the sulfur content remained more or less constant around 0.85 percent. The ash content of the lignite declined from around 11 percent in 1992 to 10 percent in 2001 but the sulfur content increased from 0.6 percent in 1992 to 0.66 percent in 2001. Though both the ash content and sulfur content of the fuels were substantial, emissions of Particulate Matter have been already controlled since 1990, the removal efficiency of Particulate Matter increased from 97 percent in 1990 to 99.5 percent in 2001. Due to the abatement measures in place and the heights of the stacks of the power plants.[54]

Coal fired power stations usually have abatement equipment such as cyclones or electrostatic filters which remove, on average, 99.5 percent of particulate emissions. Therefore, only the fine fraction, which can be transported by wind, is expected to contribute to an increase in ambient levels of suspended particulate matter in other areas.

In **Kazakhstan** as well as **Hungary**, abatement measures for PM have been implemented since early 1990s with removal efficiencies of around 96 percent going up to 99 percent at the end of the 1990s.

Stack heights: In the former USSR there were a number of state norms and rules that regulated height and design of the chimneys of the power plants. Mainly those rules and norms were quite closed to the Western regulation documents of that time.

There were national sanitary regulations that predetermined guidance values of concentrations of different pollutants (like nitrogen oxides (NOx), carbon monoxide (CO), particulate matter (PM), volatile organic compounds (VOC), and heavy metals) for practical design use.

The minimum release height (stack height) calculated with the help of an atmospheric dispersion model based on acceptable emission amounts and concentrations allowed for the power industry. The model was statistical and calculated ambient concentrations of the pollutants at ground level over a year based on emission rates and meteorological data, including a rose of wind for the region the power plant to be built The Ekibastuz Power Plant in Kazakhstan that uses coal as a fuel, for instance, has two stacks of 330 m each. Other known stack heights in Russia and Ukraine range from 250 meters to 1,370 meters.[55]

---

54. Power plants have high stacks due to the size of their boilers, the boiler needs to be maintained with underpressure in order for the poisonous gases to leave through the stack rather than in the power plants itself. The higher the level of dirty fuels (high sulfur and ash content) and the greater the size of the boiler, the higher the stack needs to be to maintain the norms for pollutants from stationary sources.
55. Berezovskaya Power Plant #1 in Russia, which has the highest chimney.

ANNEX I

Changes in Generation Mix in the Past Decade

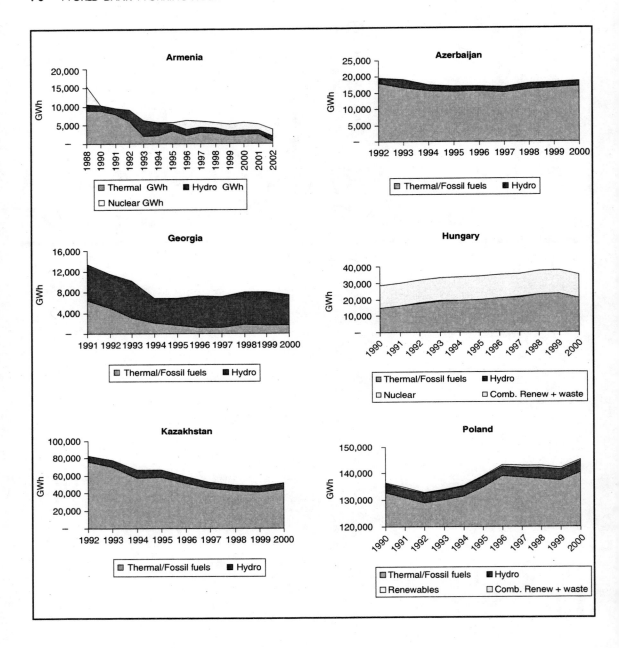

Annex J

Proposed Energy Issues to be Addressed and Sample Questions in LSMS/HBS Surveys

Fuel use

- For each energy use (heating, lighting, cooking) consumption expenditure, quantity used and price information for primary fuels, secondary fuels and other fuels. When different units are used, appropriate conversion factors need to be determined;
- Clear distinction between collected and purchased fuel wood.
- The size of the LPG container and the refill frequency to obtain quantities of LPG
- Location where fuel wood is collected and/or purchased from, average size of the collected fuel wood batch. Type of fuel wood collected/purchased, and means to collect fuel wood with. Frequency of collection.

Equipment

- What equipment do households have for burning these fuels?
- Which do they use?
- Does the fire place or equipment have a chimney?

Location of fire

- Does the household cook indoors or outdoors?
- Does the location of the cooking change with seasons?
- Does the house have a separate cooking area?
- Occasional measurements of indoor air pollutants where problems are expected

REFERENCES

Banerjee, S.G., and M.C. Munger. 2004. "Move to markets? An empirical analysis of privatization in developing countries." *Journal of International Development* 16:213–240.

Bhatnagar, B. 2001. "Filipino report card on pro-poor services." Report No. 22181-PH, Environment and Social Development Sector Unit, East Asia and Pacific, World Bank.

Birdsall, N., and J. Nellis. 2003. "Winners and Losers: Assessing the distributional impact of privatization." *World Development* 31(10):1617–1633.

Bourdaire, J.M. 1999. "Empowering the End-User: Market Reform Lessons from IEA Countries." In *Energy After the Financial Crises. Energy Development Report 1999*. World Bank.

Bucknall, J., and G. Hughes. 2000. Poland—*Complying with EU Environmental Legislation*. World Bank Technical Paper No. 454. Washington, D.C.: World Bank.

Bucknall, J. 1999. "Poland—Program for air protection in Silesia project." Project Information Document. Washington, D.C.: World Bank.

Cornille, J., and S. Frankhauser. 2002. "The energy intensity of transition countries." Working Paper 72, European Bank for Reconstruction and Development.

Dalmazzo, A., and G. de Blasio. 2003. "Resources and incentives to reform." IMF Staff Papers 50(2). Washington, D.C.: International Monetary Fund.

Esanov, A., M. Raiser, and W. Buiter. "Nature's Blessing or Nature's Curse: The Political Economy of Transition in Resource Based Economies." Working Paper 65, European Bank of Reconstruction and Development.

Estache, A., V. Foster, and Q. Wodon. 2001. *Accounting for poverty in infrastructure reform*. World Bank Institute. Washington, D.C.: World Bank.

Estache, A., A. Gomez-Lobo, and D. Leipziger, 2001. "Utility privatization and the poor: Lessons and evidence from Latin America." *World Development* 29(7):1179–1198.

European Bank for Reconstruction and Development. 2001. "Transition Report 2001: Energy in Transition."

European Environmental Agency. Europe's environment, the third assessment, air quality in Europe. Environmental Assessment Report No. 10. 2003.

Freinkman, L., G. Gyulumyan, and A. Kyurumyan. 2003. *Quasi-Fiscal Activities, Hidden Government Subsidies, and Fiscal Adjustment in Armenia.* World Bank Working Paper No. 16. Washington, D.C.: World Bank.

Freund, C.L., and C.I. Wallich, 1995. "Raising household energy prices in Poland: Who gains? Who loses." Policy Research Working Paper 1495. Washington, D.C.: World Bank.

Galal, A., L. Jones, P. Tandon, and I. Vogelsang. 1994. *Welfare consequences of selling public enterprises: An empirical analysis.* World Bank and Oxford University Press.

Guasch, J.L. 2004. "Granting and renegotiating infrastructure concessions—avoiding the pitfalls." Processed. World Bank.

Hellman, J.S. 1998. "Winners Take All: The Politics of Partial Reform in Post-Communist Transitions." *World Politics* 50(2).

Hope, E. and B. Singh, 1995. "Energy price increases in developing countries." Policy Research Paper 1442. World Bank.

International Monetary Fund. 2001a. "Republic of Armenia: Recent Economic Developments and Selected Issues." IMF Country Report 01/78.

———. 2001b. "Republic of Moldova: Recent Economic Developments and Selected Issues." IMF Country Report No. 01/22.

———. 2001c. "Republic of Georgia: Recent Economic Developments and Selected Issues." IMF Country Report 01/211.

International Energy Agency. Various years. Energy Balances of OECD- and non-OECD countries.

Katyshev, S. and G. Mandrovskaya. 2003. "Social and Environmental Impact of Electricity Reform in Kazakhstan." Processed. ECSSD, World Bank.

Kennedy, D. 1996. "Competition in the Power Sector of Transition Economies." Working Paper 41, European Bank of Reconstruction and Development.

———. 2003. "Power Sector Regulatory Reform in Transition Economies: Progress and Lessons Learned." Working Paper 78, European Bank of Reconstruction and Development.

Klytchnikova, I. 2002. "Analyzing the Welfare Effect of Electric Utilities Privatization in the Presence of Rationing. The Case of Eastern and Central Europe and Central Asia." Processed. University of Maryland, College Park.

———. 2003 "Deforestation, poverty and household use of fuel wood in the Caucasus region. A household level analysis of Azerbaijan" Mimeo. University of Maryland, College Park.

Komives, K. D. Whittington, and X. Wu. 2001. "Infrastructure coverage and the poor: A global perspective." Policy Research Working Paper 2551. World Bank.

Krishnaswamy, V. and G. Stuggins. 2003. *Private Participation in the Power Sector in Europe and Central Asia, Lessons from the Last Decade.* World Bank Working Paper No. 8. Washington, D.C.: World Bank.

Lampietti, J.A., A.A. Kolb, S. Gulyani, and V. Avenesyan. 2001. *Utility pricing and the poor: Lessons from Armenia.* World Bank Technical Paper No. 497. Washington DC.

Lampietti, J.A. and A. Meyer. 2002. *Coping with the Cold: Heating Strategies for Eastern Europe and Central Asia's Urban Poor.* World Bank Technical Paper No. 529, Washington DC.

Lampietti, J.A. and B. Kropp. 2002. "Climbing down the energy ladder? Household energy trends in Eastern Europe and Central Asia." Processed. ECSSD, World Bank.

Lampietti, J.A., H. Gonzalez, E. Hamilton, and M. Wilson. 2003. *Revisiting Reform in the Energy Sector, Lessons from Georgia.* World Bank Working Paper No. 21. Washington, D.C.: World Bank.

Lengyel, L. 2003. "Status of power sector in Hungary." Processed. Eurocorp Commerz Ltd., ECSSD, World Bank.

Lieberman, I., M. Gobbo, A. Sukiasyan, S.L. Travers, and J.R.D. Welch. 2003. "Privatization Practice Note: Europe and Central Asia Region." Processed. World Bank.

Lvovsky, K. 2000. "Environmental Costs of Fossil Fuels." World Bank Environment Department Papers No. 78, Pollution Management Series.

Markandya, A., M. Jayawardena, and R. Sharma. 2001. "The Impact of Infrastructure Investments on Measured Poverty." A Viewpoint Note. Unpublished manuscript. World Bank.

McKenzie, D. and D. Mookherjee. 2002. "Distributive impact of privatization in Latin America: An overview of evidence from four countries." *Economia* (Spring 2003)161–218.

Nadiradze, N. 2003. "Study of Social and Environmental Impacts of Electricity Reform in Georgia." Processed. ECSSD, World Bank.

Nersisyan, E. and A. Marjanyan. 2003. "Armenia Consultants Report on Power Sector and the Environment." Processed. ECSSD, World Bank.

Paul, S. 1994. "Does voice matter? For public accountability, yes." Policy Research Working Paper 1388. Washington, D.C.: World Bank.

Paul, S. 1998. "Making voice work: the report card on Bangalore's public service." Policy Research Working Paper 1921. Washington, D.C.: World Bank.

Petri, M., G. Taube, and A. Tsyvinski. 2002. "Energy Sector Quasi-Fiscal Activities in the Countries of the Former Soviet Union." IMF Working Paper WP/02/60. Washington, D.C.

Rodrik, D. 1994. "The Rush to Free Trade in the Developing World: Why So Late? Why Now? Will It Last?" In S. Haggard and S. Webb, eds., *Voting for Reform: Democracy, Political Liberalization, and Economic Adjustment.* New York: Oxford University Press.

Ryszard, G. 2003. "Status of power sector in Poland." Processed. Agency for Energy Market, ECSSD, World Bank.

Saavalainen, T.O., and J. ten Berge. 2003. "Energy Conditionality in Poor CIS Countries." Processed. Washington, D.C.: International Monetary Fund.

Sachs, J. and A. Warner. 1995. "Natural Resource Abundance and Economic Growth." NBER Working Paper No. 5398.

Shirley, M. and P. Walsh. 2000. "Public vs. private ownership: the current state of the debate." Policy Research Working Paper 2420. Washington, D.C.

Smith, K. 2000. "National burden of disease in India from indoor air pollution." School of Public Health, University of California, Berkeley. Contribution to special series of Inaugural Articles by members of National Academy of Sciences. 2000.

Srinivasan, P.V. and B.S. Reddy. 1996. "Electricity Demand Management Through Pricing: Scope and Options." *International Journal of Global Energy Issues* 8(5/6).

United Nations Environment Program. 2002. "Caucasus Environment Outlook."

United Nations Secretariat. 2002. World Population Prospects: The 2002 Revision. Population Division of the Department of Economic and Social Affairs of the United Nations Secretariat.

———. 2003. World Urbanization Prospects: The 2001 Revision.

UNECE/FAO 2000. Forest resources of Europe, CIS, North America, Australia, Japan and New Zealand: contribution to the global Forest Resources Assessment 2000. Geneva Timber and Forest Study Papers 17. New York and Geneva: United Nations.

Valiyev, V. 2003. "State of Azerbaijan Republic Electric Power Industry (1990–2002)." Shems Energy Ltd.

Vashakmadze, E. and V. Kvekvetsia. 2000. "Georgian Power Sector Deficit. Sector Note." Processed. ECSSD, World Bank.

Wodon, Q, M.I. Ajwad, J. Baker, R. Jayasuriya, C. Siaens, and J.P. Tre. 2003. Poverty and public spending in Latin America. Processed. World Bank.

World Bank. 1993a. "Armenia: Energy sector Review." Washington, D.C.

———. 1993b. "Energy Efficiency and Conservation in the Developing World: The World Bank's Role." A World Bank Policy Paper. Washington, D.C.

———. 1993. "Poland–Energy sector restructuring program, volume 1: Main report." Washington, D.C.

———. 1996. "Republic of Moldova: Energy Project." Staff Appraisal Report. Washington, D.C.

———. 1997. "Georgia: Power rehabilitation project" Staff Appraisal Report. Washington, D.C.

———. 1999. *Privatization of Power and Natural Gas Industries in Hungary and Kazakhstan*. World Bank Technical Paper No. 451. Washington, D.C.

———. 2000a. "Kazakhstan Public Expenditure Review—I, II, and III." Report No. 20489-KZ, Poverty Reduction and Economic Management Unit, Europe and Central Asia.

———. 2000b. *Making transition work for everyone: Poverty and inequality in Europe and Central Asia*. Washington DC.

———. 2002a. "Description of the Existing Power Networks in Armenia–Bank Mission aide memoires." Unprocessed. ECSIE, Washington, D.C.

———. 2002b. "Georgia Public Expenditure Review." Report No. 22913-GE, Poverty Reduction and Economic Management Unit, Europe and Central Asia.

———. 2002c. "Private sector development in the electric power sector: A joint OED/OEG/OEU Review of the World Bank's assistance in the 1990s." Processed.

———. 2003a. *World Development Indicators*. CD-ROM.

———. 2003b. "Armenia Public Expenditure Review." Report No. 24434-AM, Poverty Reduction and Economic Management Unit, Europe and Central Asia.

———. 2003c. "Azerbaijan Public Expenditure Review." Report No. 25233-AZ, Poverty Reduction and Economic Management Unit, Europe and Central Asia.

———. 2003d. "Moldova Public Economic Management Review." Report No. 25423-MD, Poverty Reduction and Economic Management Unit, Europe and Central Asia.

———. 2003e. "Poland—Toward a fiscal framework for growth, A public expenditure and institutional review." Report No. 25033-POL, Poverty Reduction and Economic Management Unit, Europe and Central Asia.

———. 2003f. "ECSPF Private Sector Strategy Paper." Processed.

———. 2004. "World Development Report—Making Services Work for Poor People" Washington, D.C.

World Health Organization. 1989. "Management and control of the Environment".

———. 2002. "Reducing risk, promoting healthy life" World Health Report.